VOLUME	EDITOR-IN-CHIEF	PAGES	
36	N. J. LEONARD	120	*Out of print*
37	JAMES CASON	109	*Out of print*
38	JOHN C. SHEEHAN	120	*Out of print*
39	THE LATE MAX TISHLER	114	*Out of print*

Collective Vol. IV A revised edition of Annual Volumes 30–39
NORMAN RABJOHN, *Editor-in-Chief*   1036

40	MELVIN S. NEWMAN	114	*Out of print*
41	JOHN D. ROBERTS	118	*Out of print*
42	VIRGIL BOEKELHEIDE	118	*Out of print*
43	B. C. MCKUSICK	124	*Out of print*
44	THE LATE WILLIAM E. PARHAM	131	*Out of print*
45	WILLIAM G. DAUBEN	118	*Out of print*
46	E. J. COREY	146	*Out of print*
47	WILLIAM D. EMMONS	140	*Out of print*
48	PETER YATES	164	*Out of print*
49	KENNETH B. WIBERG	124	*Out of print*

Collective Vol. V A revised edition of Annual Volumes 40–49
HENRY E. BAUMGARTEN, *Editor-in-Chief*   1234

Cumulative Indices to Collective Volumes, I, II, III, IV, V
RALPH L. and THE LATE RACHEL
H. SHRINER, *Editors*

50	RONALD BRESLOW	136	*Out of print*
51	RICHARD E. BENSON	209	*Out of print*
52	HERBERT O. HOUSE	192	*Out of print*
53	ARNOLD BROSSI	193	*Out of print*
54	ROBERT E. IRELAND	155	*Out of print*
55	SATORU MASAMUNE	150	*Out of print*
56	GEORGE H. BÜCHI	144	*Out of print*
57	CARL R. JOHNSON	135	*Out of print*
58	THE LATE WILLIAM A. SHEPPARD	216	*Out of print*
59	ROBERT M. COATES	267	*Out of print*

Collective Vol. VI A revised edition of Annual Volumes 50–59
WAYLAND E. NOLAND, *Editor-in-Chief*   1208

60	ORVILLE L. CHAPMAN	140	*Out of print*
61	THE LATE ROBERT V. STEVENS	165	*Out of print*
62	MARTIN F. SEMMELHACK	269	*Out of print*
63	GABRIEL SAUCY	291	*Out of print*
64	ANDREW S. KENDE	308	
65	EDWIN VEDEJS	278	
66	CLAYTON H. HEATHCOCK	265	
67	BRUCE E. SMART	289	

Collective Volumes, Cumulative Indices, and Annual Volumes 64–6
& Sons, Inc.

ORGANIC SYNTHESES

ORGANIC SYNTHESES

AN ANNUAL PUBLICATION OF SATISFACTORY
METHODS FOR THE PREPARATION
OF ORGANIC CHEMICALS

VOLUME 67
1989

BOARD OF EDITORS

BRUCE E. SMART, *Editor-in-Chief*

DAVID L. COFFEN	LEO A. PAQUETTE
CLAYTON H. HEATHCOCK	K. BARRY SHARPLESS
ALBERT I. MEYERS	JAMES D. WHITE
RYOJI NOYORI	EKKEHARD WINTERFELDT
LARRY E. OVERMAN	

THEODORA W. GREENE, *Assistant Editor*
JEREMIAH P. FREEMAN, *Secretary to the Board*
DEPARTMENT OF CHEMISTRY, *University of Notre Dame,*
NOTRE DAME, *Indiana 46556*

ADVISORY BOARD

RICHARD T. ARNOLD	ALBERT ESCHENMOSER	CHARLES C. PRICE
HENRY E. BAUMGARTEN	IAN FLEMING	NORMAN RABJOHN
RICHARD E. BENSON	E. C. HORNING	JOHN D. ROBERTS
VIRGIL BOEKELHEIDE	HERBERT O. HOUSE	GABRIEL SAUCY
RONALD BRESLOW	ROBERT E. IRELAND	R. S. SCHREIBER
ARNOLD BROSSI	CARL R. JOHNSON	DIETER SEEBACH
GEORGE H. BÜCHI	WILLIAM S. JOHNSON	MARTIN F. SEMMELHACK
T. L. CAIRNS	ANDREW S. KENDE	JOHN C. SHEEHAN
JAMES CASON	N. J. LEONARD	RALPH L. SHRINER
ORVILLE L. CHAPMAN	B. C. McKUSICK	H. R. SNYDER
ROBERT M. COATES	SATORU MASAMUNE	EDWIN VEDEJS
E. J. COREY	WATARU NAGATA	KENNETH B. WIBERG
WILLIAM G. DAUBEN	MELVIN S. NEWMAN	PETER YATES
WILLIAM D. EMMONS	WAYLAND E. NOLAND	

FORMER MEMBERS OF THE BOARD, NOW DECEASED

ROGER ADAMS	ARTHUR C. COPE	OLIVER KAMM
HOMER ADKINS	NATHAN L. DRAKE	C. S. MARVEL
C. F. H. ALLEN	L. F. FIESER	C. R. NOLLER
WERNER E. BACHMANN	R. C. FUSON	W. E. PARHAM
A. H. BLATT	HENRY GILMAN	WILLIAM A. SHEPPARD
WALLACE H. CAROTHERS	CLIFF S. HAMILTON	LEE IRVIN SMITH
H. T. CLARKE	W. W. HARTMAN	ROBERT V. STEVENS
J. B. CONANT	JOHN R. JOHNSON	MAX TISHLER
		FRANK C. WHITMORE

JOHN WILEY & SONS
NEW YORK · CHICHESTER · BRISBANE · TORONTO · SINGAPORE

Published by John Wiley & Sons, Inc.

Copyright © 1989 by Organic Syntheses, Inc.

All rights reserved. Published simultaneously in Canada.

Reproduction or translation of any part of this work beyond that permitted by Section 107 or 108 of the 1976 United States Copyright Act without the permission of the copyright owner is unlawful.

"John Wiley & Sons, Inc. is pleased to publish this volume of Organic Syntheses on behalf of Organic Syntheses, Inc. Although Organic Syntheses, Inc. has assured us that each preparation contained in this volume has been checked in an independent laboratory and that any hazards that were uncovered are clearly set forth in the write-up of each preparation, John Wiley & Sons, Inc. does not warrant the preparations against any safety hazards and assumes no liability with respect to the use of the preparations."

Library of Congress Catalog Card Number: 21-17747
ISBN 0-471-51379-2

Printed in the United States of America

10 9 8 7 6 5 4 3 2 1

NOTICE

With Volume 62, the Editors of *Organic Syntheses* began a new presentation and distribution policy to shorten the time between submission and appearance of an accepted procedure. The soft cover edition of this volume is produced by a rapid and inexpensive process, and is sent at no charge to members of the Organic Divisions of the American and French Chemical Society, The Perkin Division of the Royal Society of Chemistry, and The Society of Synthetic Organic Chemistry, Japan. The soft cover edition is intended as the personal copy of the owner and is not for library use. A hard cover edition is published by John Wiley and Sons Inc. in the traditional format, and differs in content primarily in the inclusion of an index. The hard cover edition is intended primarily for library collections and is available for purchase through the publisher. Annual Volumes 60–64 are being incorporated into a new five-year version of the collective volumes of *Organic Syntheses* which will appear as *Collective Volume Seven* in the traditional hard cover format. It will be available for purchase from the publishers. The Editors hope that the new *Collective Volume* series, appearing twice as frequently as the previous decennial volumes, will provide a permanent and timely edition of the procedures for personal and institutional libraries. The Editors welcome comments and suggestions from users concerning the new editions.

NOMENCLATURE

Both common and systematic names of compounds are used throughout this volume, depending on which the Editor-in-Chief felt was more appropriate. The *Chemical Abstracts* indexing name for each title compound, if it differs from the title name, is given as a subtitle. Systematic *Chemical Abstracts* nomenclature, used in both the 9th and 10th Collective Indexes for the title compound and a selection of other compounds mentioned in the procedure, is provided in an appendix at the end of each preparation. Registry numbers, which are useful in computer searching and identification, are also provided in these appendixes. Whenever two names are concurrently in use and one name is the correct *Chemical Abstracts* name, that name is adopted. For example, both diethyl ether and ethyl ether are normally used. Since ethyl ether is the established *Chemical Abstracts* name for the 8th Collective Index, it has been used in this volume. The 9th Collective Index name is 1,1'-oxybisethane, which the Editors consider too cumbersome.

SUBMISSION OF PREPARATIONS

Organic Syntheses welcomes and encourages submission of experimental procedures which lead to compounds of wide interest or which illustrate important new developments in methodology. The Editorial Board will consider proposals in outline format as shown below, and will request full experimental details for those proposals which are of sufficient interest. Submissions which are longer than three steps from commercial sources or from existing *Organic Syntheses* procedures will be accepted only in unusual circumstances.

Organic Syntheses Proposal Format

1. Authors
2. Literature reference or enclose preprint if available.
3. Proposed sequence
4. Best current alternative(s)
5. a. Proposed scale, final product:

b. Overall yield:
  c. Method of isolation and purification:
  d. Purity of product (%):
  e. How determined?
6. Any unusual apparatus or experimental technique:
7. Any hazards?
8. Source of starting material?
9. Utility of method or usefulness of product.

Submit to: Dr. Jeremiah P. Freeman, Secretary
             Department of Chemistry
             University of Notre Dame
             Notre Dame, IN 46556

Proposals will be evaluated in outline form, again after submission of full experimental details and discussion, and, finally by checking experimental procedures. A form that details the preparation of a complete procedure (Notice to Submitters) may be obtained from the Secretary.

Additions, corrections, and improvements to the preparations previously published are welcomed; these should be directed to the Secretary. However, checking of such improvements will only be undertaken when new methodology is involved. Substantially improved procedures have been included in the Collective Volumes in place of a previously published procedure.

ACKNOWLEDGMENT

Organic Syntheses wishes to acknowledge the contributions of E. I. du Pont de Nemours and Co., Inc., Hoffmann-La Roche, Inc., and the Rohm and Haas Co. to the success of this enterprise through their support, in the form of time and expenses, of members of the Boards of Directors and Editors.

PREFACE

This volume contains 30 reliable preparations that demonstrate new general synthetic methods or provide specific compounds of broad interest to synthetic chemists. The synthesis of chiral molecules and the use of organometallic reagents are again emphasized in this annual volume. The first part comprises procedures for making various optically pure materials, including some binaphthyl derivatives that have proved to be among the most useful chiral auxiliaries for asymmetric synthesis. It opens with Jacques and Fouquey's preparation of **1,1'-BINAPH-THYL-2,2'-DIYL HYDROGEN PHOSPHATE** enantiomers, followed by Truesdale's procedure for converting one of them to **(R)-(+)-1,1'-BINAPHTHALENE-2,2'-DIOL.** Next is Noyori's route to both optically pure enantiomers of **BINAP.** The effectiveness of BINAP as a ligand for metal-catalyzed asymmetric reactions is amply illustrated by the accompanying procedure for making **(R)-(−)-N,N-DIETHYL-(E)-CITRONELLALENAMINE AND (R)-(+)-CITRONELLAL** via the Rh(I)-catalyzed asymmetric isomerization of **N,N-DIETHYLGERAN-YLAMINE** or **N,N-DIETHYLNERYLAMINE.** A new chiral auxiliary for activating secondary amines toward metallation and alkylation is **(S)-N,N-DIMETHYL-N'-(1-tert-BUTOXY-3-METHYL-2-BUTYL)FOR-MAMIDINE,** and its use in asymmetric methylation is nicely demonstrated in the preparation of **(−)-SALSOLIDINE.** An improved synthesis of the chiral synthon **N-tert-BUTOXYCARBONYL-L-LEUCINAL** is described next, and the final procedure in this part, **CONDENSATION OF (−)-DIMENTHYL SUCCINATE DIANION WITH 1,ω-DIHAL-IDES,** presents a general method for making optically active trans-cycloalkane-1,2-dicarboxylic acids.

The next 15 procedures illustrate the diverse uses of metallic reagents to promote synthetically useful transformations. The first five involve Pd catalysis, beginning with the preparations of **ETHYL (E)-4-(4-NITRO-PHENYL)-4-OXO-2-BUTENOATE** and **ETHYL 5-OXO-6-METHYL-6-HEPTENOATE** which entail Pd(0)-catalyzed couplings of acid chlorides with organotin and organozinc reagents, respectively. Next are examples of **1,4-FUNCTIONALIZATION OF 1,3-DIENES VIA PALLADIUM-CATALYZED CHLOROACETOXYLATION AND ALLYLIC AMINATION** and **PALLADIUM(0)-CATALYZED syn-1,4-ADDITION OF CARBOXYLIC ACIDS TO CYCLOPENTADIENE MONOEPOXIDE,** followed by the two-step preparation of **4,4-DI-**

METHYL-2-CYCLOPENTEN-1-ONE, which employs a Pd(II)-Cu(I) catalyzed oxidation of 2,2-dimethyl-4-pentenal to 2,2-dimethyl-4-oxopentanal in the first step.

Organosilicon reagents play a key role in each of the next five procedures. The synthesis of **2-METHYL-2-UNDECENE FROM ETHYL DECANOATE** demonstrates the use of α-(diphenylmethylsilyl)esters as vinyl dication equivalents, and **N-BENZYL-N-METHOXYMETHYL-N-(TRIMETHYLSILYL)METHYLAMINE** is shown in the next procedure to be an effective azomethine ylide equivalent. The reaction of trimethylsilylenol ethers with hexacarbonyl(propargylium)dicobalt salts provides a general method for α-propargylation of ketones, which is exemplified by the synthesis of **2-(1-METHYL-2-PROPYNYL)CYCLOHEXANONE** from cyclohexanone. Finally, Paquette makes use of an aluminum chloride promoted reaction between bis(trimethylsilyl)acetylene and p-toluenesulfonyl chloride in his synthesis of **ETHYNYL p-TOLYL SULFONE,** which is a useful Michael acceptor and acetylene synthon in Diels-Alder reactions.

This section of the volume concludes with procedures involving the elements Se, Ni, Al, Ti, and Rh. Paquette's method for dienophile activation via selenosulfonation is illustrated by the preparation of **1-(BENZENESULFONYL)CYCLOPENTENE,** and the accompanying preparation of **BICYCLO[4.3.0]NON-1-EN-4-ONE** gives an example of the versatile use of such α,β-unsaturated sulfone dienophiles. Nickel(II) is used as a catalyst in the procedure to make **ETHYL α-(HEXAHYDROAZEPINYLIDENE-2)-ACETATE FROM O-METHYLCAPROLACTIM AND MELDRUM'S ACID,** which is a specific example of a general route to cyclic β-enamino methyl and ethyl esters. The next procedure uses a combination of tri-isobutyl aluminum and diiodomethane to effect the **SELECTIVE CYCLOPROPANATION OF (S)-(−)-PERILLYL ALCOHOL.** This new method is notable for its regioselectivity, which differs from that of the traditional Simmons-Smith cyclopropanation reaction or its modifications. Seebach's synthesis of **3'-NITRO-1-PHENYLETHANOL BY ADDITION OF METHYLTRIISOPROPOXYTITANIUM TO m-NITROBENZALDEHYDE** elegantly illustrates the selectivity of this organotitanium reagent for carbonyl groups. This section concludes with an interesting synthesis of **N-ACETYL-N-PHENYLHYDROXYLAMINE VIA CATALYTIC TRANSFER HYDROGENATION OF NITROBENZENE USING HYDRAZINE AND RHODIUM ON CARBON.**

Four of the last five entries in this volume are convenient procedures to make functionalized molecules that are useful precursors to more complex structures: **METHYL 7-HYDROXYHEPT-5-YNOATE** and **4-METHOXY-3-PENTEN-2-ONE,** both of which, for example, are used in making prostaglandins; **3-HYDROXY-1-CYCLOHEXENE-1-CARBOXALDEHYDE;** and **(E)-2-(1-PROPENYL)CYCLOBUTANONE.** The final procedure describes a remarkably selective **4-CHLORINATION OF ELECTRON-RICH BENZENOID COMPOUNDS** by N-chlorodialkyl amines.

I wish to thank Dr. Theodora W. Greene for her conscientious editorial assistance and especially Professor Jeremiah P. Freeman, our indefatigable Secretary, for his help in preparing the text and structures for this volume. All of the structures were drawn using the ChemDraw™ program. I also am grateful to my colleagues on the Editorial Board who made my tenure such an enjoyable and rewarding experience.

BRUCE E. SMART

Wilmington, Delaware
July 1988

CARL SHIPP MARVEL
September 11, 1894–January 4, 1988

Carl S. Marvel had a spectacular career of 72 years in organic chemistry. From 1920 to 1961 he was on the staff of the University of Illinois and from the date of his first retirement through 1987, he was a faculty member at the University of Arizona. He consulted for nearly 60 years for the DuPont Company. He was a dominant figure in American organic chemistry and has been recognized as the "father" of synthetic polymer chemistry. His early contributions to *Organic Syntheses* are easily recognizable by perusal of Collective Volume I of this series, which contained the preparations that appeared in the first nine annual volumes, was published in 1932, and was reprinted in 1941. Nearly 20% of the 264 preps in that collective volume were either submitted by Marvel or checked by him. It was Marvel's experience with the "summer preps" group, or—more formally—"Organic Chemical Manufactures", at the

University of Illinois starting in 1916, plus his provision and checking of preps for *Organic Syntheses,* that gave him wide-ranging synthetic experience.

Born on a farm near Waynesville, Illinois, Carl Marvel attended a one-room grammar school and then Waynesville Academy, where he thrived on Latin and Greek. He was introduced to chemistry as a freshman at Illinois Wesleyan University in 1911. An uncle who had been a high school teacher advised his nephew to take this subject if he expected to be a farmer, since the next generation of farmers was going to require scientific knowledge to get the most out of their work. At Illinois Wesleyan, Carl Marvel found enjoyment in organic chemistry and was delighted to learn from his professor, Alfred W. Homberger, that he could be paid to study further, by means of a $250 scholarship, at the University of Illinois. His graduate education in 1915 started with an overload of course work, including four lab courses, in order to "catch up". When he was not studying, he worked late at night in the laboratory. As a result, he slept as late as possible but still got to the breakfast table before the dining room door closed at 7:30 a.m. His student colleagues decided that was the only time he ever hurried, and they nicknamed him "Speed". A nickname was appropriate to his friendly spirit, but it causes us to smile because it was really an accurate moniker indicative of his chemical thinking, his human insight, his fishing and bird-watching prowess, and the alacrity with which he found out how he could help a student or colleague with a chemical or personal problem.

We all know from the literature after 1920 and from his many award citations what followed in Speed Marvel's research when he joined the faculty of the University of Illinois, starting with synthesis as the initial motivation and moving boldly into areas of rearrangements, free radical chemistry and magnetic susceptibility, hydrogen bonding, stereoisomerism, structure of organo-mercury and phosphorus compounds, and—most important of all—polymers. While he always felt, and often reminded his colleagues, that the essential product of academic research was the students, he also taught that the best graduate training was to be achieved, along with possible national prestige, by work on essential problems. He believed that there was no such thing as a dead end to a worthwhile problem—delays and detours and retracing of steps, indeed, but no dead end. His research on polymers started with the peroxide-initiated reaction of sulfur dioxide with olefins. He then became interested in the basic mechanism of vinyl polymerization and how vinyl units went together to form a polymer. His research contributions to synthetic rubber, initiated during the second world war under the auspices of the

National Defense Research Council and lasting into the mid-50's, together with the findings of his technical intelligence mission after the war, had a lasting impact on the American rubber industry. Marvel's association with the U.S. Air Force research program began toward the end of his first career at the University of Illinois and continued throughout his entire second career at the University of Arizona. During a 30-year period he was the principal contributor to the Air Force program on high temperature polymer synthesis. His basic research led to the commercialization of polybenzimidazole (PBI) which, because of its exceptional resistance to fire, is used in the suits of astronauts and fire fighters.

Speed Marvel was a founder of the High Polymer Forum that became the Division of Polymer Chemistry of the American Chemical Society, of which he became Chairman in 1950–1951. During his 74-year membership in the ACS, he held just about every elective office possible up to and including the presidency in 1945. The all-purpose meeting room of the ACS building in Washington, D.C. is designated "Marvel Hall" to indicate the esteem with which the ACS held Speed Marvel and the Society's gratitude for the leadership he provided in raising the funds that made the building possible. During his long career he gave unselfishly of his time on a variety of committees and editorial boards, as he did also for his two Universities. Both Arizona and Illinois have annual Marvel Lectureships and Marvel Scholarships. In 1984, the University of Arizona renamed the chemistry laboratory building where he worked the Carl S. Marvel Laboratories of Chemistry.

Honors in steady stream were awarded to Marvel during his career, culminating in the Distinguished Service Award from the U.S. Air Force Materials Laboratory and the National Medal of Science. Other awards included the Nichols, Gibbs, Priestley, and Perkin Medals and election to the Plastics Hall of Fame.

Speed will long be remembered by every chemist who came in contact with him. His students, colleagues, and fellow members of the Boards of *Organic Syntheses* will particularly cherish his memory. He has left us a legacy we can all appreciate.

NELSON J. LEONARD

April 21, 1988

CONTENTS

J. Jacques and C. Fouquey	1	ENANTIOMERIC (S)-(+)- AND (R)-(-)-1,1'-BINAPHTHYL-2,2'-DIYL HYDROGEN PHOSPHATE
Larry K. Truesdale	13	(R)-(+)-1,1'-BINAPHTHALENE-2,2'-DIOL
Hidemasa Takaya, Susumu Akutagawa, and Ryoji Noyori	20	(R)-(+)- AND (S)-(-)-2,2'-BIS(DIPHENYLPHOSPHINO)-1,1'-BINAPHTHYL (BINAP)
Kazuhide Tani, Tsuneaki Yamagata, Sei Otsuka, Hidenori Kumobayashi, and Susumu Akutagawa	33	(R)-(-)-N,N-DIETHYL-(E)-CITRONELLALENAMINE AND (R)-(+)-CITRONELLAL VIA ISOMERIZATION OF N,N-DIETHYLGERANYLAMINE OR N,N-DIETHYLNERYLAMINE
Kunihiko Takabe, Takashi Katagiri, Juntaro Tanaka, Tsutomu Fujita, Shoji Watanabe, and Kyoichi Suga	44	ADDITION OF DIALKYLAMINES TO MYRCENE: N,N-DIETHYLGERANYLAMINE
Kunihiko Takabe, Takashi Yamada, Takao Katagiri, and Juntaro Tanaka	48	TELOMERIZATION OF ISOPRENE WITH DIALKYLAMINE: N,N-DIETHYLNERYLAMINE
Daniel A. Dickman, Michael Boes, and Albert I. Meyers	52	(S)-N,N-DIMETHYL-N'(1-tert-BUTOXY-3-METHYL-2-BUTYL)FORMAMIDINE
Albert I. Meyers, Michael Boes, and Daniel A. Dickman	60	(-)-SALSOLIDINE
O. P. Goel, U. Krolls, M. Stier, and S. Kesten	69	N-tert-BUTOXYCARBONYL-L-LEUCINAL
Kyoji Furuta, Kiyoshi Iwanaga, and Hisashi Yamamoto	76	CONDENSATION OF (-)-DIMENTHYL SUCCINATE DIANION WITH $1,\omega$-DIHALIDES: (+)-(1S,2S)-CYCLOPROPANE-1,2-DICARBOXYLIC ACID
A. F. Renaldo, J. W. Labadie, and J. K. Stille	86	PALLADIUM-CATALYZED COUPLING OF ACID CHLORIDES WITH ORGANOTIN REAGENTS: ETHYL (E)-4-(4-NITROPHENYL)-4-OXO-2-BUTENOATE
Yoshinao Tamaru, Hirofumi Ochiai, Tatsuya Nakamura, and Zen-ichi Yoshida	98	ETHYL 5-OXO-6-METHYL-6-HEPTENOATE FROM METHACRYLOYL CHLORIDE AND ETHYL 4-IODOBUTYRATE

Authors	Page	Title
J. E. Nyström, T. Rein, and J. E. Bäckvall	105	1,4-FUNCTIONALIZATION OF 1,3-DIENES VIA PALLADIUM-CATALYZED CHLORO-ACETOXYLATION AND ALLYLIC AMINATION: 1-ACETOXY-4-DIETHYLAMINO-2-BUTENE AND 1-ACETOXY-4-BENZYLAMINO-2-BUTENE
Donald R. Deardorff and David C. Myles	114	PALLADIUM(O)-CATALYZED syn-1,4-ADDITION OF CARBOXYLIC ACIDS TO CYCLOPENTADIENE MONOEPOXIDE: cis-3-ACETOXY-5-HYDROXYCYCLOPENT-1-ENE
David Pauley, Frank Anderson, and Tomas Hudlicky	121	4,4-DIMETHYL-2-CYCLOPENTEN-1-ONE
Gerald L. Larson, Ingrid Montes de Lopez-Cepero, and Luis Rodriguez Mieles	125	α-DIPHENYLMETHYLSILYLATION OF ESTER ENOLATES: 2-METHYL-2-UNDECENE FROM ETHYL DECANOATE
Albert Padwa and William Dent	133	N-BENZYL-N-METHOXYMETHYL-N-(TRIMETHYLSILYL)METHYLAMINE AS AN AZOMETHINE YLIDE EQUIVALENT: 2,6-DIOXO-1-PHENYL-4-BENZYL-1,4-DIAZABICYCLO[3.3.0]OCTANE
Valsamma Varghese, Manasi Saha, and Kenneth M. Nicholas	141	ALKYLATIONS USING HEXACARBONYL-(PROPARGYLIUM)DICOBALT SALTS: 2-(1-METHYL-2-PROPYNYL)-CYCLOHEXANONE
Liladhar Waykole and Leo A. Paquette	149	ETHYNYL p-TOLYL SULFONE
Ho-Shen Lin, Michael J. Coghlan, and Leo A. Paquette	157	DIENOPHILE ACTIVATION VIA SELENOSULFONATION: 1-(BENZENESULFONYL)CYCLOPENTENE
Ho-Shen Lin and Leo A. Paquette	163	REDUCTIVE ANNULATION OF VINYL SULFONES: BICYCLO[4.3.0]NON-1-EN-4-ONE
J. P. Celerier, E. Deloisy-Marchalant, G. Lhommet, and P. Maitte	170	ETHYL α-(HEXAHYDROAZEPINYLIDENE-2)-ACETATE FROM O-METHYLCAPROLACTIM AND MELDRUM'S ACID
Keiji Maruoka, Soichi Sakane, and Hisashi Yamamoto	176	SELECTIVE CYCLOPROPANATION OF (S)-(-)-PERILLYL ALCOHOL: 1-HYDROXYMETHYL-4-(1-METHYLCYCLO-PROPYL)-1-CYCLOHEXENE
René Imwinkelried and Dieter Seebach	180	3'-NITRO-1-PHENYLETHANOL BY ADDITION OF METHYLTRIISOPROPOXY-TITANIUM TO m-NITROBENZALDEHYDE

Authors	Page	Title
P. W. Oxley, B. M. Adger, M. J. Sasse, and M. A. Forth	187	N-ACETYL-N-PHENYLHYDROXYLAMINE VIA CATALYTIC TRANSFER HYDROGENATION OF NITROBENZENE USING HYDRAZINE AND RHODIUM ON CARBON
Guy Casy, John W. Patterson, and Richard J. K. Taylor	193	METHYL 7-HYDROXYHEPT-5-YNOATE
George A. Kraus, Michael E. Krolski, and James Sy	202	4-METHOXY-3-PENTEN-2-ONE
H. L. Rigby, M. Neveu, D. Pauley, B. C. Ranu, and T. Hudlicky	205	3-HYDROXY-1-CYCLOHEXENE-1-CARBOXALDEHYDE
Scott A. Miller and Robert C. Gadwood	210	SYNTHESIS OF CYCLOBUTANONES VIA 1-BROMO-1-ETHOXYCYCLOPROPANE: (E)-2-(1-PROPENYL)CYCLOBUTANONE
John R. Lindsay Smith, Linda C. McKeer, and Jonathan M. Taylor	222	4-CHLORINATION OF ELECTRON-RICH BENZENOID COMPOUNDS: 2,4-DICHLOROMETHOXYBENZENE
Unchecked Procedures	232	
Cumulative Author Index for Volumes 65, 66, and 67	236	
Cumulative Subject Index for Volumes 65, 66, and 67	239	

ORGANIC SYNTHESES

ENANTIOMERIC (S)-(+)- AND (R)-(-)-1,1'-BINAPHTHYL-2,2'-DIYL HYDROGEN PHOSPHATE

(Dinaphtho[2,1-d:1',2'-f][1,3,2]dioxaphosphepin, 4-hydroxy-, 4-oxide)

A.

B.

Submitted by J. Jacques and C. Fouquey.[1]
Checked by P. R. Carlier and K. Barry Sharpless.

1. Procedure

Caution! Part A of this procedure should be carried out in an efficient hood to avoid exposure to noxious vapors (pyridine, phosphorus oxychloride).

A. *(±)-Binaphthylphosphoric (BNP) acid.* A 1-L, three-necked flask, fitted with a magnetic stirring bar, pressure-equalizing dropping funnel, reflux condenser topped by a calcium chloride drying tube, and a thermometer, is charged with 450 mL of pyridine and, while stirring, with 100 g (0.35 mol) of (±)-1,1'-bi-2-naphthol (Note 1).

To this stirred suspension, 73.6 g (0.48 mol) of freshly distilled phosphorus oxychloride is added dropwise, whereupon the temperature rises to about 80°C, most of the binaphthol dissolves, and pyridine hydrochloride crystals form. Complete dissolution is achieved by heating to 90°C. The stirred solution is allowed to cool to 50-60°C (crystallization occurs at about 85°C). To the stirred suspension, 40 mL of water is added dropwise (*Caution, exothermic reaction!*), which raises the temperature to the boiling point (ca. 118°C). The resulting solution, cooled to about 60°C, is transferred to a 1-L dropping funnel and the flask is rinsed with pyridine (2 x 20 mL). The solution and rinse are combined and added dropwise with vigorous stirring to 900 mL of 6 N hydrochloric acid (Note 2), which gives a precipitate of pyridine-solvated binaphthylphosphoric (BNP) acid (Note 3). This crude product is collected by suction filtration. The wet cake is transferred to a 2-L, large-necked flask and stirred with 300 mL of 6 N hydrochloric acid. The suspension is heated to boiling (possible foaming!) and immediately cooled. The solid is thoroughly filtered by suction, washed twice with 20 mL of water (Note 4), and air-dried to afford 114-119 g (94-99%) of (±)-binaphthylphosphoric acid. This compound, which decomposes without melting at about 300°C (Note 5), is pure enough to be resolved. Analytical crystalline samples can be obtained from ethanol.

B. *(S)-(+)- and (R)-(-)-BNP acid.* In a 2-L flask 95.2 g (0.27 mol) of racemic binaphthylphosphoric (BNP) acid and 80.4 g (0.27 mol) of (+)-cinchonine (Note 6) are dissolved in 985 mL of hot methanol (Note 7). To the hot (65°C) solution is added 420 mL of hot water via a dropping funnel over the course of 20 min. *During the addition the solution is vigorously stirred and maintained at 65-70°C. At the end of the addition, the flask is transferred to another (cool) stirring plate. Crystallization starts at approximately 60°C, and stirring is maintained until the solution has reached room temperature* (Note 8). The crystals are collected, washed with a 2:1 methanol-water mixture (3 x 45 mL) and air dried to afford 76.6 g of salt consisting of 91% p salt [(+)-acid, (+)-base] and 9% n salt [(-)-acid, (+)-base], $[\alpha]_{546}^{25}$ +424° (methanol, *c* 0.99) (Notes 9 and 10).

(S)-(+)-BNP acid. A 2-L, three-necked flask, equipped with addition funnel, reflux condenser, magnetic stirring bar, and thermometer, is charged with 76.6 g of the above salt and 500 mL of ethanol. The salt is dissolved by heating to reflux, and 570 mL of 6 N hydrochloric acid is added with vigorous stirring over the course of 30 min. The temperature is maintained between 75-80°C during the addition, and the acid begins to precipitate. Once the addition is complete, the solution is allowed to cool without stirring to room temperature. The solid is collected, washed with water (5 x 90 mL), and air-dried to afford 26.7 g of (S)-(+)-BNP acid, $[\alpha]_{546}^{25}$ +712° (methanol, *c* 0.98) (Note 11). The yield based on enantiomer present in the racemate is 56%. The product is free from contamination by cinchonine, as shown by ^{1}H NMR, (Me$_2$SO, 250 MHz) and elemental analysis. HPLC analysis of the methyl ester derivative employing a chiral stationary phase (Note 12) shows the acid to be greater

than 99.4% ee. Partially resolved samples, recovered by adding water to the filtrates, may be purified by crystallization from ethanol or by digestion in hot methanol (Note 13).

(R)-(-)-BNP acid. The filtrate from the initial crystallization of the cinchonine salt is evaporated nearly to dryness to give 107 g of crude salt, $[\alpha]_{546}^{25}$ -113° (methanol, *c* 0.95), consisting of approximately 81% n salt and 19% p salt (Notes 14 and 10). A 2-L, three-necked flask, equipped with addition funnel, reflux condenser, magnetic stirring bar, and thermometer is charged with 107 g of the crude salt and 700 mL of ethanol. The salt is dissolved by heating to reflux, and 790 mL of 6 N hydrchloric acid is added with vigorous stirring over the course of 30 min. The temperature is maintained between 75-80°C during the addition, and the acid begins to precipitate. Once the addition is complete, the solution is allowed to cool without stirring to room temperature. The solid is collected, washed with water (5 x 100 mL), and air-dried to afford 23.3 g of (-)-BNP acid, $[\alpha]_{546}^{25}$ -717° (methanol, *c* 1.00) (Note 11). The yield based on enantiomer present in the racemate is 49%. The product is free from contamination by cinchonine, as shown by ^{1}H NMR (Me$_2$SO, 250 MHz) and elemental analysis. HPLC analysis of the methyl ester derivative employing a chiral stationary phase (Note 12) shows the acid to be 100.0% ee.

2. Notes

1. Commercial dry pyridine, stored over 4 Å molecular sieves, was used without further purification. The 1,1'-bi-2-naphthol is commercially available from Aldrich Chemical Company, Inc. The submitters prepared it by oxidizing a hot aqueous suspension of commercial 98% pure 2-naphthol (Merck-

Schuchardt) with ferric chloride[2] to obtain crude colored binaphthol (80-90% yield). Unless this material is purified and decolorized by successive crystallization and digestion in hot toluene, the color will be retained in the binaphthylphosphoric acid (60-65% overall yield).

2. The reverse addition of 6 N hydrochloric acid to the pyridine solution results in the formation of a thick and syrupy precipitate, which prevents stirring.

3. Regardless of the conditions of precipitation or crystallization, a polymorphic solvate is obtained which consists of 2 BNP acid:1 pyridine:1 H_2O, according to elemental analyses. Pyridine peaks are apparent in the ^1H NMR spectrum (δ 8.71 and 8.78 in d_6-DMSO). Desolvation occurs at 210-230°C on the Kofler bench or by heating to reflux in 6 N hydrochloric acid.

4. The solubility of BNP acid in water is about 2 g/L at 20°C.

5. The BNP acid is polymorphic. A metastable form, identical with the enantiomers and therefore a conglomerate,[3] was sometimes obtained when working above 40°C; the usual stable form is a racemic compound.[3] The IR spectrum (Nujol, cm^{-1}) of the racemate is as follows: 950 (strong), 1025 (strong, broad), 1185 (medium), 1200 (strong), 1220 (strong); of conglomerate: 1050 (strong, broad), 1200 (medium), 1230 (strong), 1255 (medium).

6. Commercial (+)-cinchonine (Aldrich Chemical Company, Inc.), $[\alpha]_D^{25}$ +228° (ethanol, c 0.5), was used without further purification.

7. The checkers observed coloration of this solution and stirred it with 10 g of Norit activated carbon, followed by filtration through a Celite pad. The pad was washed with hot methanol (2 x 50 mL). The submitters did not report any coloration.

8. In order to achieve an efficient resolution it is imperative that the addition of water be even and slow; otherwise premature precipitation or oiling of the cinchonine salt may occur. Likewise, stirring must be maintained during the cooling period to achieve high yields and to avoid the formation of oils. The salt is collected as soon as the mixture cools to room temperature, because the more soluble salt deposits as an oil on standing. The submitters suggest that the yield may be improved by carrying out the crystallization in a cold bath until the solution reaches room temperature. It should be noted that the checkers did not observe crystallization until a packet of seed crystals was opened in their laboratory.

9. The submitters obtained 78.5 g of salt composed of 97% p and 3% n salts, $[\alpha]_{546}^{25}$ +471° (methanol, c 0.9).

10. The p and n salts, prepared from the pure (+)- or (-)-acids and cinchonine and crystallized from methanol-ethyl acetate and methanol-acetone-ethyl acetate, respectively, exhibit the following rotations (± 3%) in methanol:

$$[\alpha]_\lambda^{25}$$

	589 nm	578 nm	546 nm	436 nm	
p salt	+409°	+428	+492	+890	(c 0.7)
n salt	-211	-222	-256	-474	(c 0.8)

11. Two crystallizations from ethanol did not change the optical rotations of (+)- and (-)-BNP acid, which are as follows:

$[\alpha]_\lambda^{25}$

	589 nm	578 nm	546 nm	436 nm	365 nm
Methanol	595° ± 7	624 ± 7	720 ± 8	1328 ± 15	2050 ± 25
Ethanol	574 ± 16	602 ± 17	694 ± 20	1267 ± 25	1828 ± 40

Both enantiomers decompose without melting above 300°C. The solid state IR spectra of the enantiomers and the racemate (conglomerate) are identical. The checkers obtained a rotation of -705° at 546 nm.

12. The methyl ester derivatives were prepared by treating BNP acid with diazomethane in methanol-ether. A Regis Pirkle Type 1-A preparative column (25 cm x 10 mm I.D.) was used and the conditions were as follows: 10% 2-propanol/hexanes, 8.0 mL min^{-1}, detector at 284 nm. The (R)-(-) enantiomer is eluted first and the peaks are well separated (α = 1.24).[4]

13. BNP acid, racemate and enantiomers are sparingly soluble in water and organic solvents, except alcohols. Their solubilities at 25 ± 0.5°C in methanol and 95% ethanol, expressed in g/100 mL of solvent and g/100 mL of solution (in brackets) are given below:

	Racemate	Enantiomer
Ethanol	10.3 ± 0.5 (11.5)	5.7 ± 0.2 (6.7)
Methanol	2.5 ± 0.1 (3.1)	2.1 ± 0.1 (2.6)

As shown by these data, the racemate is approximately twice as soluble as the enantiomers in ethanol, whereas in methanol, the solubilities are nearly the same. This is because the racemate forms a crystalline compound (solvate) with methanol as shown by IR and NMR (Me$_2$SO) spectra, and by elemental analysis. The racemate dissolves more rapidly in refluxing methanol than do the enantiomers. Accordingly, partially resolved samples having 80-90% ee can be conveniently purified by merely digesting them for 10-15 min in refluxing methanol. In general, partially resolved BNP acid can be purified by crystallization from ethanol. In this case, the dissolution rate is particularly slow and the desired solution is obtained by using solvent in excess, then concentrating the solution.

14. The submitters obtained 103 g of material consisting of about 85% n and 15% p salts, $[\alpha]_{546}^{25}$ -150° (methanol, c 0.8).

3. Discussion

Marschalk[5] first prepared BNP acid by the action of phosphorus oxychloride on binaphthol without any solvent, followed by hydrolysis of the isolated acid chloride. The procedure herein described was only briefly mentioned, without any experimental details, in a paper describing the resolution of BNP acid and its use as a resolving agent.[6a]

The BNP acid has also been prepared by Cram and co-workers[7] from binaphthol and phosphorus oxychloride, although under quite different conditions, which involved isolation of the intermediate acid chloride, its hydrolysis by aqueous tetrahydrofuran, and extraction of BNP acid with ethyl acetate. In our hands, this extraction could not be carried out without using larger volumes of solvent and water than those reported. More recently, it has also been prepared by Japanese authors who used a slightly modified method.[8]

The resolution of BNP acid by crystallizing the (+)-acid-cinchonine salt, then the (-)-acid-cinchonidine salt, was first mentioned in a short paper and subsequently described in a patent.[6a]

A modification of this procedure was used by Cram, et al.[7] who likewise obtained the (+)-BNP acid via its cinchonine salt, but they isolated the (-) enantiomer directly from the more soluble salt, without using cinchonidine, in 59% and 46% respective yields. Both enantiomers, precipitated by 6 N hydrochloric acid from their cinchonine salt solutions, had to be purified by successive digestions with hot 6 N hydrochloric acid and water in order to decompose any remaining cinchonine salt. These purifications are avoided by using the procedure described herein.

BNP acid has also been resolved via its strychnine salt.[6b]

Some physical properties of BNP acid have been studied (triplet state circular dichroism,[9a] luminescence, photoracemization.[9b]).

Derivatives have been prepared: methyl esters (enantiomers and racemate)[6] and D-glucopyranosyl ester.[10]

Enantiomeric (S)-(+)- and (R)-(-)-BNP acids are useful resolving agents, which give well crystallized, easily separated salts with a variety of amines. They have been used in the preparation of the enantiomers of biologically- and therapeutically-active compounds, such as α-difluoromethyl-α-aminovaleric acid,[11] cephalosporin,[12] dibenzothiepin,[13] benzodiazepine derivatives[14] and 3-hydroxyphenyl-N-propylpiperidine.[15] BNP acid also has been used for the direct resolution of underivatized o-tyrosine.[16]

(+)-BNP acid, linked to silica gel, has been used in high performance liquid chromatographic resolution of helicenes.[17]

Reduction of (S)-(+)- and (R)-(-)-BNP methyl esters[6a] or acids[7] by lithium aluminum hydride, or by Red-Al (this volume, p. 13) yields (S)-(-)- and (R)-(+)-binaphthol, respectively. This is, at present, the most convenient access to optically active binaphthols, used by Cram and co-workers[18] to prepare macrocyclic polyethers and by Japanese authors[8] in asymmetric synthesis of cyclic binaphthyl-esters.

Racemic and optically active BNP acids (as well as binaphthols) are also available commercially from Aldrich Chemical Company, Inc.

1. Laboratoire de Chimie des Interactions Moléculaires, Equipe de Recherche associée au CNRS, E.R. 285, Collège de France, 11, Place Marcelin Berthelot, 75231 Paris Cedex 05, France.
2. Pummerer, R.; Prell, E.; Rieche, A. *Ber.* **1926**, *59B*, 2159. See also Rieche, A.; Jungholt, K.; Frühwald, E. *Ber.* **1931**, *64B*, 578, Note 12.
3. Leclercq, M.; Collet, A.; Jacques, J. *Tetrahedron* **1976**, *32*, 821; Jacques, J.; Collet, A.; Wilen, S. H. "Enantiomers, Racemates and Resolutions", Wiley: New York, 1981; p. 18.
4. Pirkle, W. H.; Schreiner, J. L. *J. Org. Chem.* **1981**, *46*, 4988-4991.
5. Marschalk, C. *Bull. Soc. Chim.* **1928**, *43*, 1388.
6. a) Jacques, J.; Fouquey, C.; Viterbo, R. *Tetrahedron Lett.* **1971**, 4617; Viterbo, R.; Jacques, J. Ger. Offen. Patent 2 212 660, 1972; *Chem. Abstr.* **1973**, *78*, 43129b; b) Hoyano, Y. Y.; Pincock, R. E. *Can. J. Chem.* **1980**, *58*, 134.
7. Kyba, E. P.; Gokel, G. W.; de Jong, F.; Koga, K.; Sousa, L. R.; Siegel, M. G.; Kaplan, L.; Sogah, G. D. Y.; Cram, D. J. *J. Org. Chem.* **1977**, *42*, 4173.

8. Miyano, S.; Tobita, M.; Hashimoto, H. *Bull. Chem. Soc. Jpn.* **1981**, *54* 3522.
9. (a) Tétreau, C.; Lavalette, D. *Nouv. J. Chim.* **1980**, *4*, 423; (b) Tétreau, C.; Lavalette, D.; Cabaret, D.; Geraghty, N.; Welvart, Z. *Nouv. J. Chim.* **1982**, 6, 461.
10. Schmidt, R. R.; Stumpp, M.; Michel, J. *Tetrahedron Lett.* **1982**, *23*, 405.
11. Bey, P.; Vevert, J. P.; Van Dorsselaer, V.; Kolb, M. *J. Org. Chem.* **1979**, *44*, 2732 and patents to Metcalf, B. W.; Jung, M. Belg. Patent 868 593, 1978, *Chem. Abstr.*, **1979**, *90*, 187336n; Bey, P.; Jung, M. Belg. Patent 881 210, 1980; *Chem. Abstr.* **1981**, *94*, 145343q; Bey, P.; Jung, M. Belg. Patent 881 208 1980, *Chem. Abstr.* **1981**, *95*, 103310s; Bey, P.; Jung, M. Belg. Patent 881 209, 1980, *Chem. Abstr.* **1981**, *95*, 103311t; Bey, P.; Jung, M. U.S. Patent 4 309 442, 1982 ; *Chem. Abstr.* **1982**, *96*, 205402m; Bey, P.; Jung, M. U.S. Patent 4 330 559, 1982; *Chem. Abstr.* **1982**, *97*, 144373z.
12. Edwards, M. L. Belg. Patent 874 662, 1979; *Chem. Abstr.* **1980**, *92*, 128941z.
13. Kyburz, E.; Aschwanden, W. Ger. Offen Patent 2 625 258, 1976; *Chem. Abstr.* **1978**, *89*, 180040g; Kyburz, E.; Aschwarden, W. Ger. Offen. Patent 2 625 258, 1976; *Chem. Abstr.* **1979**, *90*, 23113m.
14. Werner, W.; Jungstand, W.; Gutsche, W.; Wohlrabe, K. Ger. (East) D.D. Patent 149 527, 1981; *Chem. Abstr.* **1982**, *96*, 69053u.
15. Arnold, W.; Daly, J. J.; Imhof, R.; Kyburz, E. *Tetrahedron Lett.* **1983**, *24*, 343.
16. Garnier-Suillerot, A.; Albertini, J. P.; Collet, A.; Faury, L.; Pastor, J. M.; Tosi, L. *J. Chem. Soc., Dalton Trans.* **1981**, 2544.
17. Mikes, F.; Boshart, G. *J. Chem. Soc., Chem. Commun.* **1978**, 173; Mikes, F.; Boshart, G. *J. Chromatogr.* **1978**, *149*, 455.

18. Cram, D. J.; Helgeson, R. C.; Peacock, S. C.; Kaplan, L. J.; Domeier, L. A.; Moreau, P.; Koga, K.; Mayer, J. M.; Chao, Y.; Siegel, M. G.; Hoffman, D. H.; Sogah, G. D. Y. *J. Org. Chem.* **1978**, *43*, 1930; Cram, D. J. U.S. Patent 4 043 979, 1977; *Chem. Abstr.* **1978**, *89*, 109618w.

Appendix
Chemical Abstracts Nomenclature (Collective Index Number);
(Registry Number)

Cinchonine (8); 9S-Cinchonan-9-ol (9); (118-10-5)

Cinchonidine (8); 8α,9R-Cinchonan-9-ol (9); (485-71-2)

S(+) BNP acid; S(+) Dinaphtho[2,1-d:1'2'-f][1,3,2]dioxaphosphepin, 4-hydroxy-4-oxide (9); (35193-64-7). Compound with 9S-cinchonan-9-ol (9); (39749-50-3)

R(-) BNP acid; R(-) Dinaphtho[2,1-d:1'2'-f][1,3,2]dioxaphosphepin, 4-hydroxy-4-oxide (9); (39648-67-4). Compound with 8α,9R-cinchonan-9-ol (9); (40481-36-5)

(R)-(+)-1,1'-BINAPHTHALENE-2,2'-DIOL

([1,1'-Binaphthalene]-2,2'-diol, (R)-)

Submitted by Larry K. Truesdale.[1]
Checked by Georg W. Schröder and K. Barry Sharpless.

1. Procedure

A. (R)-(-)-Methyl 1,1'-binaphthyl-2,2'-diylphosphate. A 200-mL, three-necked flask equipped with a magnetic stirrer and a gas bubbler is charged with 20.0 g (57.4 mmol) of (R)-(-)-1,1'-binaphthyl-2,2'-diyl hydrogen phosphate (Note 1), 40 mL of N,N-dimethylacetamide, and 10.0 mL (105.7 mmol) of dimethyl sulfate (Note 2). The resulting pale yellow oil is then treated cautiously in small portions with 10.4 g (123.8 mmol) of sodium bicarbonate.

The addition causes gas evolution and foaming. Foaming subsides after ca. 20 min, and the resulting turbid yellow solution is stirred overnight at room temperature (Note 3).

The reaction mixture is poured into a mixture of 300 mL of toluene and 100 mL of ethyl acetate. The resulting milky solution is washed twice with 100-mL portions of deionized water, twice with 100-mL portions of brine, and then dried over anhydrous sodium sulfate. The drying agent is removed by filtration and the filtrate is concentrated on a rotary evaporator under reduced pressure in a 50°C bath. The semi-solid residue is slurried in 55 mL of ether and then collected by filtration on a sintered glass funnel. After the solid is washed four times with 10 mL ether, it is set aside and the filtrate is evaporated to dryness. The residue is slurried in 10 mL of ether. The solid is collected on a glass frit and washed twice with 10 mL of ether. Both solids are combined and dried under reduced pressure to give 16.4 g (45.2 mmol) (79%) of phosphate as an off-white powder, mp 215-217°C; $[\alpha]_D^{25}$ -526.8° (THF, c 1.16) (Note 4).

B. *Crude (R)-(+)-1,1'-binaphthalene-2,2'-diol. [Gases evolved in this step create a stench and are toxic. The use of an efficient fume hood is imperative (Note 5).]*

A 2-L, three-necked flask equipped with a Y-tube, magnetic stirrer, dropping funnel topped with a gas bubbler, thermometer, and drying tube is charged with 14.7 g (40.6 mmol) of crude (R)-(-)-methyl 1,1'-binaphthyl-2,2'-diyl phosphate and 350 mL of dry toluene. The mixture is stirred under nitrogen and heated with a steam bath until dissolution occurs (44°C). The solution is then cooled in an ice-water bath to 10°C. The cooling bath is removed and a solution of 27.0 mL (91.8 mmol) of Red-Al (Note 6) in 35 mL of toluene is added from the dropping funnel over a 90-min period. The mixture turns yellow, evolves a gas, and heats up to 26°C. Gas evolution ceases 20 min after the addition is completed. TLC analysis (silica gel plate developed with ethyl acetate) indicates that the reaction is complete.

The entire reaction mixture is poured into 430 mL of 10% hydrochloric acid (mild exothermic reaction). The organic layer is separated and washed with another 430-mL portion of 10% hydrochloric acid. After further washes with 150 mL of brine and 160 mL of deionized water, all aqueous layers are combined and back-extracted with a 550-mL portion of 3:2 toluene:methanol and two 300-mL portions of toluene. The combined organic layers are dried over anhydrous sodium sulfate, and filtered. The solvent is removed on a rotary evaporator under reduced pressure in a 45°C bath to give 11.4 g (98%) of crude product as a bright yellow microcrystalline solid, mp 203-205°C.

C. Recrystallization of (R)-(+)-1,1'-binaphthalene-2,2'-diol. A 200-mL flask with a reflux condenser is charged with 11.4 g of crude (R)-(+)-1,1'-binaphthalene-2,2'-diol and 1.5 g of Norit A, and refluxed for 5 min. The hot solution is filtered through a pad of 5 g of Celite on a glass frit. The Celite and Norit A from the frit are slurried together with 1.5 g of Norit A in 70 mL of toluene, refluxed for 5 min, and filtered hot through a pad of 5 g of Celite on a glass frit. The Celite is washed with 70 mL of hot toluene. The combined filtrates are warmed to 50°C to dissolve the precipitated crystals and filtered once again while warm through a pad of 5 g of Celite on a glass frit, which is then washed with 40 mL of hot toluene. The filtrate is freed from solvent on a rotary evaporator and the residue is recrystallized from 75 mL of toluene with stirring. The mixture is stirred overnight at ambient temperature. The resulting suspension is filtered and the filter cake is washed with two 10-mL portions of toluene. The recrystallized product is then dried under reduced pressure (high vacuum) to give 9.62 g of white, microcrystalline material in the first crop. This material melts at 207-209°C, $[\alpha]_D^{25}$ +33.6° (THF, c 1.11) (Notes 7 and 8).

Concentration of the filtrate and toluene washings under reduced pressure to ca. 11 mL affords, after cooling, collecting, washing three times with 3 mL of toluene, and drying the solid material, a second crop of product which weighs 1.04 g. This crop is an off-white microcrystalline powder, mp 204-206°C and $[\alpha]_D^{25}$ +33.6° (THF, c 1.12) (Note 9).

The two crops (10.66 g) represent 94% recovery on recrystallization. The yield of recrystallized product is 92% in the reductive cleavage of the phosphate ester and 73% overall.

2. Notes

1. The starting material was prepared by the method described in the companion procedure of Fouquey and Jacques (*Org. Synth.* **1988**, *67*, 1) and had $[\alpha]_D^{25}$ -704.7 (MeOH, c 1.0).

2. Dimethyl sulfate was obtained from Aldrich Chemical Company, Inc. It is highly toxic and carcinogenic, and it should be handled only in a well-ventilated hood.

3. The reaction was complete by TLC (silica-gel plate) analysis. The consumption of starting material and the formation of product can be monitored using a 5:2 (v:v) mixture of ethyl acetate:hexane as the solvent.

4. The enantiomeric excess of this product was determined to be >99.5% using a chiral stationary phase HPLC (preparative Regis Pirkle Type 1-A, 10 x 250 mm I.D., 7.5 mL/min flow rate, 1000 psi pressure, 10% 2-propanol in hexane, detector at 284 nm). The R-(-)-enantiomer is eluted first and the peaks are well separated.[2] Another batch of phosphate ($[\alpha]_D^{25}$ -507.7°C, THF, c 1.17) was shown to have 96.5% ee using the same conditions.

The same reaction was also checked on a slightly smaller scale, using 18.8 g (54.0 mmol) of (R)-(-)-1,1'-binaphthyl-2,2'-diyl phosphate. The yield was 81%. The submitters report a yield of 80% for the reaction run on a 1.6 mole scale.

5. The phosphorus hydride which evolves is highly toxic and creates a stench. Therefore the checkers recommend that this reaction as well as the workup be performed in an efficient fume hood. The checkers found it advantageous to scrub the phosphorus hydride which formed in a trap (shown below) by bubbling the evolving gases through a solution of 250 mL of bleach (5% sodium hypochlorite in water). During the reaction a constant flow of nitrogen was applied to avoid contamination of the reaction flask with wet gases from the traps.

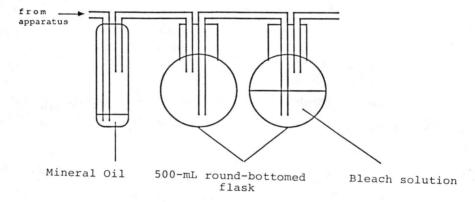

6. Red-Al is the Aldrich Chemical Company, Inc., brand of sodium bis(2-methoxyethoxy)aluminum hydride in 3.4 M toluene solution.

7. The enantiomeric excess of this product was determined to be >99.5% using the same conditions as mentioned in Note 4. The peaks are well separated and the R-(+)-enantiomer is eluted second.[2] Another batch obtained from reduction of phosphate having 96.5% ee had a mp of 207-209°C and $[\alpha]_D^{25}$

+33.5° (THF, c 1.12). This material had an enantiomeric excess of >99.5%, determined under the same conditions as described in Note 4. The submitters report $[\alpha]_D^{25}$ +34.7° (THF, c 1.035) for material with mp 207-209°C [lit.[5] mp 206.5-207.5°C, $[\alpha]_D^{25}$ +34.3° (THF, c 1.1)].

8. The product has the following spectral properties: ^{1}H NMR (1:1 CDCl$_3$:DMSO-d$_6$) δ: 7.04 (d, 2 H, J = 8.8), 7.20-7.35 (m, 4 H), 7.40 (d, 2 H, J = 8.8), 7.92 (d, 2 H, J = 8.8), 9.21 (s, 2 H).

9. The enantiomeric excess of this product was determined to be >99.5% using the same conditions as mentioned in Note 4. Another batch obtained from the reduction of phosphate having 96.5% ee had a mp of 200-204°C and $[\alpha]_D^{25}$ +30.7° (THF, c 1.13). This (second crop) material had an enantiomeric excess of 86.5%, determined under the same conditions as described in Note 4. The submitters report $[\alpha]_D^{25}$ +33.5° (THF, c 0.775) for material with mp 204-206°C.

3. Discussion

Enantiomerically pure 1,1'-binaphthalene-2,2'-diols are used in various types of asymmetric syntheses, for example, as chiral auxiliaries in a method for the asymmetric reduction of ketones.[2]

The previously published method[4,5] for the liberation of the diol from the resolved 1,1'-binaphthyl-2,2'-diyl hydrogen phosphate entailed esterification with diazomethane and reductive cleavage with lithium aluminum hydride. The procedure presented here is felt to be safer in that it circumvents the hazards associated with using diazomethane and lithium aluminum hydride on a large scale.

1. Chemistry Research Department, Hoffmann-La Roche Inc., Nutley, NJ 07110.
2. Pirkle, W. H.; Schreiner, J. L. *J. Org. Chem.* **1981**, *46*, 4988.
3. Fieser, M.; Danheiser, R. L.; Roush, W. in "Fieser and Fieser's Reagents for Organic Synthesis", Wiley-Interscience: New York, 1981; Vol, 9, pp. 169-170.
4. Jacques, J.; Fouquey, C.; Viterbo, R. *Tetrahedron Lett.* **1971**, 4617.
5. Kyba, E. P.; Gokel, G. W.; de Jong, F.; Koga, K.; Sousa, L. R.; Siegel, M. G.; Kaplan, L.; Sogah, G. D. Y.; Cram, D. J. *J. Org. Chem.* **1977**, *42*, 4173.

Appendix
Chemical Abstracts Nomenclature (Collective Index Number);
(Registry Number)

(±)-Binaphthol; [1,1'-Binaphthalene]-2,2'-diol (8,9); (41024-90-2), (602-09-5)

R-(+)-Binaphthol; [1,1'-Binaphthalene]-2,2'-diol (8,9); (18531-94-7)

S-(-)-Binaphthol; [1,1'-Binaphthalene]-2,2'-diol (8,9); (18531-99-2)

(R)-(-)-Methyl 1,1'-binaphthyl-2,2'-diyl phosphate: Dinaphtho[2,1-d:1',2'-f][1,3,2]dioxaphosphepin, 4-methoxy-, 4-oxide, (R)- (11); (86334-02-3)

(R)-(-)-1,1'-Binaphthyl-2,2'-diyl hydrogen phosphate: Dinaphtho[2,1-d:1',2'-f][1,3,2]dioxaphosphepin, 4-hydroxy-, 4-oxide, (R)- (9); (39648-67-4)

Sodium bis(2-methoxyethoxy)aluminum hydride: Aluminate (1-), dihydrobis(2-methoxyethanolato)-, sodium (8); Aluminate(1-), dihydrobis(2-methoxyethanolato-0,0')-, sodium (9); (22722-98-1)

(R)-(+)- AND (S)-(-)-2,2'-BIS(DIPHENYLPHOSPHINO)-1,1'-BINAPHTHYL (BINAP)

(Phosphine, [1,1'-binaphthalene]-2,2'-diylbis[diphenyl-, (R)- or (S)-)

Submitted by Hidemasa Takaya[1a], Susumu Akutagawa,[1b] and Ryoji Noyori.[1c]
Checked by Marco Cereghetti, Alain Rageot, Max Vecchi, and Gabriel Saucy.

1. Procedure

Caution! These operations, which involve toxic reagents, should be conducted in an efficient hood.

A. *2,2'-Dibromo-1,1'-binaphthyl.* A 2-L, three-necked, round-bottomed flask is equipped with an efficient mechanical stirrer, thermometer, and a dropping funnel. The flask is charged with 240 g (0.915 mol) of triphenylphosphine and 500 mL of dry acetonitrile (Note 1). Stirring is begun and the solid is dissolved by warming the flask with hot water. The solution is then cooled with an ice/water mixture and to this is added dropwise with stirring 155 g (50 mL, 0.969 mol) of bromine over a period of 1 hr. The ice/water bath is removed and 120 g (0.420 mol) of 2,2'-dihydroxy-1,1'-binaphthyl is added to the solution (Note 2). The viscous slurry is stirred at 60°C for 30 min. The flask is now fitted for a simple distillation and most of the solvent is removed by applying partial vacuum. The last trace of acetonitrile is removed at aspirator vacuum using a bath temperature of 100°C. The temperature of the resulting mass is raised carefully by means of a heating mantle (Note 3) to 240-260°C over a period of 1 hr, at which temperature an exothermic reaction occurs (Note 4), with evolution of hydrogen bromide. After the exothermic reaction subsides, the reaction mixture is further stirred at 260-270°C for 1 hr, and then the temperature is gradually raised and kept at 310-320°C for 30 min to complete the reaction. The reaction mixture, a homogenous melt, is allowed to cool to ca. 200°C with stirring and to this is added 1000 mL of Celite with stirring (Note 5). After the mixture is cooled below 70°C, it is extracted with 500 mL of a boiling 1:1 mixture of benzene and hexane. The solid material, separated by filtration through a sintered-glass funnel, is extracted further with three 200-mL portions of a boiling 1:1 mixture of benzene and hexane. The combined extracts are evaporated to give an orange-yellow viscous oil, which is dissolved in 200 mL of ethanol. The solution is left in a refrigerator for 2 days (Note 6). 2,2'-Dibromo-1,1'-binaphthyl precipitates and is collected on a sintered-glass funnel to give 90 g of the crude product. Recrystallization from ethanol affords the pure dibromide (78.0 g, 45% yield) as pale yellow, fine crystals (Note 7).

B. *(±)-2,2'-Bis(diphenylphosphinyl)-1,1'-binaphthyl [(±)-BINAPO]*. A 1-L, three-necked, round-bottomed flask is provided with a mechanical stirrer, addition funnel, thermometer, and a reflux condenser, the top of which is connected with a bubbler and an argon line by way of a three-way stopcock. The flask is flushed with argon and charged with 2.84 g (0.117 g-atom) of magnesium turnings, 50 mL of dry, degassed tetrahydrofuran (Note 8), 50 mg of iodine, and 0.5 mL of 1,2-dibromoethane. The mixture is stirred at room temperature until the color of iodine fades and evolution of ethylene ceases. The flask is placed in an oil bath, the reaction mixture is stirred and heated at 50°-70°C, and 20.0 g (50.0 mmol) of 2,2'-dibromo-1,1'-binaphthyl in 400 mL of dry, degassed toluene (Note 9) is added over a period of 3.5 hr. The mixture is stirred at 75°C for 2 hr and then cooled to 10°C. To this is added dropwise over a 20 min period a solution of 28.4 g (120 mmol) of diphenylphosphinyl chloride (Note 10) in 35 mL of toluene (Note 9) while the temperature is held at 10-15°C. After the addition is completed, the mixture is further stirred at 60°C for 2 hr, and then cooled to 15°C. To the solution is added dropwise 350 mL of 10% aqueous ammonium chloride and the mixture is stirred for another 10 min at 60°C. The organic layer is separated, washed successively with 150 mL of 10% aqueous ammonium chloride, two 150-mL portions of 1 N sodium hydroxide, and finally with two 150-mL portions of water. The toluene layer is dried for a short time over anhydrous sodium sulfate (Note 11), filtered, and concentrated under reduced pressure to give 38.8 g of a pale yellow solid. This crude product is stirred with 150 mL of boiling toluene for a few minutes and to this is added 100 mL of heptane. The mixture is allowed to stand at room temperature overnight. The solid product is separated by filtration through a sintered-glass funnel and dried at 70°C (0.05 mm) for 2 hr to give 24.5 g (75%) of (±)-BINAPO as a slightly pale yellow solid (Note 12). Concentration of the filtrate and recrystallization of the residue twice from 30-mL portions of toluene gives an additional 3.6 g (11%) of (±)-BINAPO. This product is suitable for use in Part C without further purification.

C. *Optical resolution of (±)-BINAPO.* A 2-L, round-bottomed flask is equipped with a magnetic stirrer bar and a reflux condenser. The flask is charged with 10.5 g (16.0 mmol) of racemic BINAPO and 700 mL of chloroform (Note 13). The solid is dissolved by heating at reflux temperature with stirring, followed by rapid addition of a warm solution of 6.0 g (16.0 mmol) of (-)-2,3-O-dibenzoyl-L-tartaric acid monohydrate [(-)-DBT monohydrate] (Note 14) in 460 mL of ethyl acetate (Note 13). The mixture is stirred under reflux for 2-3 min and then allowed to stand at room temperature overnight. The crystals formed are collected on a sintered-glass funnel and the filtrate is stored for recovery of (R)-(+)-BINAPO (see below). The solid product is dried at room temperature (0.05 mm) for 6 hr to give 7.2 g (89% of theory) of a 1:1 complex of (S)-BINAPO and (-)-DBT, mp 238-240°C (dec), $[\alpha]_D^{25}$ -170° (ethanol, *c* 0.503) (Note 15).

This complex (7.1 g, 7.0 mmol) is treated with 150 mL of 0.75 N aqueous sodium hydroxide and the mixture is extracted with two 150-mL portions of chloroform. The combined organic layers are washed with 100 mL of 0.75 N sodium hydroxide, water, and dried over anhydrous sodium sulfate. The drying agent is removed by filtration and the solvent is evaporated. The residue is washed with 20 mL of cold ethyl acetate to furnish 5.3 g of white solid, which is dried at 80°C (0.05 mm) overnight to give 4.6 g (100% based on the complex used) of (S)-BINAPO, mp 256-258°C, $[\alpha]_D^{25}$ -392° (benzene, c 0.530) (Note 16).

The mother liquor and the filtrate from the first resolution (see above) are combined and concentrated to dryness to give 9.0 g of solid material ((R)-BINAPO and (-)-DBT) after being dried at 80°C (0.05 mm) for 3 hr, mp 228-230°C (dec). This solid is treated with 150 mL of 0.75 N aqueous sodium hydroxide and extracted with two 150-mL portions of chloroform. The combined extract is washed with 70 mL of 0.75 N sodium hydroxide, two 100-mL portions of water, and dried over sodium sulfate. The drying agent is removed by filtration and the filtrate is evaporated to give 7.7 g of colorless solid, which is dried at 80°C overnight to afford 5.9 g (9.0 mmol) of crude (R)-BINAPO, mp 249-251°C, $[\alpha]_D^{20}$ +304° (benzene, c 0.522). This recovered (R)-BINAPO is dissolved in 350 mL of refluxing chloroform and to this is added with stirring a solution of 3.4 g (9.0 mmol) of (+)-DBT monohydrate in 280 mL of warm ethyl acetate. The mixture is stirred at reflux temperature for 5 min and then allowed to stand at room temperature overnight. The white precipitates are collected on a sintered-glass funnel, washed with two 20-mL portions of cold ethyl acetate, and dried at 70°C (0.05 mm) for 12 hr to give 7.5 g [92% yield based on the initially used (R)-BINAPO] of the (R)-BINAPO-(+)-DBT-complex, mp 235-236°C (dec), $[\alpha]_D^{25}$ +172° (ethanol, c 0.527).

This complex (7.3 g, 7.2 mmol) is treated with 200 mL of 0.75 N aqueous sodium hydroxide and extracted twice with 150-mL portions of chloroform. The combined chloroform layer is washed with 60 mL of 0.75 N aqueous hydroxide, two 100-mL portions of water, and dried over anhydrous sodium sulfate, and filtered. Evaporation of the filtrate affords 5.25 g of colorless solid which is dried at 80°C (0.05 mm) to give 4.65 g (99% yield based on the complex used) of (R)-BINAPO, mp 256-258°C, $[\alpha]_D^{25}$ +388° (benzene, c 0.514) (Note 17).

D. *Reduction of (S)-(-)-BINAPO to (S)-(-)-BINAP.* In a 300-mL, three-necked flask, fitted with a magnetic stirrer bar, thermometer, and a reflux condenser which is connected through a three-way stopcock to an argon inlet tube and a bubbler, is placed 4.5 g (6.9 mmol) of (S)-BINAPO. The flask is flushed with argon followed by the addition of 100 mL of dry, degassed xylene (Note 9), 4.2 mL of triethylamine (3.1 g, 30 mmol) (Note 9), and 3.0 mL (4.0 g, 29 mmol) of trichlorosilane (Note 13) by means of syringes. The mixture is stirred and heated at 100°C for 1 hr, at 120°C for 1 hr, and finally at refluxing temperature for 6 hr (Note 18). After the solution is cooled to room temperature, 70 mL of 30% aqueous sodium hydroxide solution is carefully added. The mixture is then stirred at 60°C until the organic and aqueous layers become clear, and it is transferred into a 300-mL separatory funnel. The organic layer is separated, and the aqueous layer is extracted with two 50-mL portions of warm toluene. The combined organic layer is washed with 70 mL of 30% sodium hydroxide solution and three 100-mL portions of water, and then dried over anhydrous sodium sulfate. The organic layer is concentrated under reduced pressure to a volume of about 15 mL and to this is added 15 mL of degassed methanol. The precipitates are collected on a sintered-glass funnel, washed with 15 mL of methanol, and dried at 80°C (0.05 mm) for 6 hr to give 4.2 g (97% yield) of (S)-BINAP as colorless solid, mp 236-238°C, $[\alpha]_D^{25}$ -223° (benzene, c 0.502) (Notes 19 and 20).

2. Notes

1. Reagent grade acetonitrile was dried over 3Å molecular sieves and then heated at reflux for several hours over calcium hydride and distilled under dry argon.

2. Commercial reagent grade 2,2'-dihydroxy-1,1'-binaphthyl from Aldrich Chemical Company, Inc. (1,1'-bi-2-naphthol) was used as obtained. It also can be prepared by the oxidative coupling of β-naphthol with ferric chloride[2] and used after recrystallization from ethanol and then benzene.

3. In order to obtain a homogeneous melt, the checkers had to raise the bath temperature to 300-335°C. They preferred a Woods metal bath.

4. Temperature should be carefully controlled. Too rapid heating can result in an uncontrollably vigorous reaction.

5. This facilitates the smooth extraction of the products.

6. This procedure removes triphenylphosphine oxide.

7. The product melts at 184-185°C (lit.[3] mp 180°C), R_f 0.50 (E. Merck Kieselgel 60 PF_{254}, 1:4 benzene-hexane).

8. Reagent grade tetrahydrofuran was distilled from sodium benzophenone ketyl under argon before use.

9. Commercial guaranteed grade solvents were distilled over finely powdered calcium hydride under argon before use.

10. Commercial reagent grade diphenylphosphinyl chloride from Aldrich Chemical Company, Inc. was used as obtained. This compound can be prepared either by oxidation of diphenylphosphinous chloride with dimethyl sulfoxide[4] or by the treatment of diphenylphosphinic acid with phosphorus pentachloride.[5]

11. Prolonged standing of the solution at room temperature may cause precipitation of (±)-BINAPO. In such a case, warm chloroform can be used to dissolve the solid.

12. The checkers obtained first crops ranging from 76 to 84%, mp 295.5-297°C. The submitters report mp 299-300°C. An analytically pure sample was obtained by recrystallization from a mixture of hexane and toluene, mp 304-306°C. One gram of (±)-BINAPO dissolves in 28 mL of boiling toluene.

13. Commercial reagent grade chemicals were used.

14. Guaranteed grade (-)-2,3-0-dibenzoyl-L-tartaric acid monohydrate and its enantiomer were purchased from Tokyo Kasei Kogyo Co., Ltd., and used without further purification.

15. The checkers had yields ranging from 69 to 90%. They determined the enantiomeric purity of the S-BINAPO component, $[\alpha]_D^{20}$ -168° (ethanol, c 0.5), to be 99.6/0.4 using a Pirkle column (Note 16). The submitters report that recrystallization from a 1:2 mixture of ethyl acetate and chloroform gave an analytically pure sample, mp 240-241°C (dec), $[\alpha]_D^{24}$ -174° (ethanol, c 0.523).

16. The checkers obtained first crops of 87% and 75.8%, mp 263-263.5°C, $[\alpha]_D^{20}$ -389° (benzene, c 0.5), and mother liquors of 11% and 13.5% respectively. These materials were analyzed on a Pirkle column (Baker bond II) with hexane/ethanol mixtures and found to have S/R ratios of 99.7/0.3 (first crop) and 93/1.7 (mother liquor). The submitters report obtaining analytically pure (S)-BINAPO by recrystallization from a mixture of hexane and toluene, mp 261-262°C, $[\alpha]_D^{24}$ -396° (benzene, c 0.467).

17. The submitters obtained analytically pure (R)-BINAPO by recrystallization from a mixture of hexane and toluene, mp 262-263°C, $[\alpha]_D^{24}$ +399° (benzene, c 0.500). See Note 16 for determination of optical purity. The checkers found an R/S ratio of 98.8/1.2 for unrecrystallized material, mp 261-261°C, $[\alpha]_D^{20}$ +379° (benzene, c 0.5).

18. During this period a white solid forms at the bottom of the reflux condenser. Use of a ground-glass joint as large as possible is recommended to avoid clogging.

19. GLC analysis (OV-101, capillary column, 5 m, 200-280°C) indicates that the product has a purity of 97%. Trace amounts of BINAPO and the monooxide of BINAP were detected by TLC analysis (E. Merck Kieselgel 60 PF$_{254}$, 1:19 methanol-CHCl$_3$); R$_f$ 0.42 (BINAPO), 0.67 (monooxide of BINAP), and 0.83 (BINAP). The submitters report that recrystallization from a 1:1 mixture of toluene and ethanol affords optically pure (S)-BINAP, mp 241-242°C, $[\alpha]_D^{25}$ -229° (benzene, c 0.312). The checkers oxidized a sample of first crop material, mp 241-242.5°C, $[\alpha]_D^{20}$ -221° (benzene, c 0.5), for Pirkle analysis (see Note 16). This gave an S/R ratio of 98.2/1.7.

20. The checkers also reduced (R)-(+)-BINAPO to (R)-(+)-BINAP by this procedure. In the best of two runs, first crop material (94.8%), mp 241-242°C, $[\alpha]_D^{20}$ +217° (benzene, c 0.5), with an R/S ratio of 99.0/0.8 was obtained.

3. Discussion

BINAP is a new type of fully aryl-substituted diphosphine with only an axial element of chirality. Optically pure BINAP was first synthesized by the optical resolution of (±)-BINAP using optically active di-μ-chlorobis[(S)-N,N-dimethyl-α-phenylethylamine-2C,N]dipalladium.[6] The phosphine is also obtained by stereospecific transformation of optically active 2,2'-dibromo-1,1'-binaphthyl.[6b,7] The procedure outlined here,[8] however, is the best preparative-scale synthesis of both enantiomers of BINAP in an optically pure state starting from easily accessible racemic 2,2'-dihydroxy-1,1'-binaphthyl. This method is applicable to various BINAP analogues.[8] The absolute configuration of (+)-BINAP was determined to be R by X-ray analysis of the complex [Rh((+)-binap)(norbornadiene)]ClO$_4$.[9] BINAP serves as an excellent ligand for the Rh(I)-catalyzed asymmetric hydrogenations of α-(acylamino)acrylic acids and esters.[6] The ligand has also been successfully applied to the Rh(I)-catalyzed asymmetric isomerization of diethylnerylamine or diethylgeranylamine into citronellal (E)-N,N-diethylenamine (this volume, p. 33).[10] This reaction is now used for commercial production of (-)-menthol. In addition, BINAP-based Ru(II) complexes[11] catalyze highly enantioselective hydrogenation of alkyl- or aryl-substituted acrylic acids,[12] enamides leading to isoquinoline alkaloids,[13] allylic and homoallylic alcohols,[14] β-keto esters,[15] other functionalized ketones,[16] etc.

1. (a) Chemical Materials Center, Institute for Molecular Science, Myodaiji, Okazaki 444, Japan. (b) Central Research Laboratory, Takasago Perfumery Co., Ltd., 5-Chome, Kamata, Ohta-ku, Tokyo 144, Japan. (c) Department of Chemistry, Nagoya University, Chikusa, Nagoya 464, Japan.
2. Pummerer, R.; Prell, E.; Rieche, A. *Ber.* **1926**, *59B*, 2159-2175.
3. Pichat, L.; Clément, J. *Bull. Soc. Chim. France* **1961**, 525-528.
4. Amonoo-Neizer, E. H.; Ray, S. K.; Shaw, R. A.; Smith, B. C. *J. Chem. Soc.* **1965**, 4296-4300.
5. Higgins, Wm. A.; Vogel, P. W.; Craig, W. G. *J. Am. Chem. Soc.* **1955**, *77*, 1864-1866.
6. (a) Miyashita, A.; Yasuda, A.; Takaya, H.; Toriumi, K.; Ito, T.; Souchi, T.; Noyori, R. *J. Am. Chem. Soc.* **1980**, *102*, 7932-7934; (b) Miyashita, A.; Takaya, H.; Souchi, T.; Noyori, R. *Tetrahedron* **1984**, *40*, 1245-1253.

7. Brown, K. J.; Berry, M. S.; Waterman, K. C.; Lingenfelter, D.; Murdoch, J. R. *J. Am. Chem. Soc.* **1984**, *106*, 4717-4723.
8. Takaya, H.; Mashima, K.; Koyano, K.; Yagi, M.; Kumobayashi, H.; Taketomi, T.; Akutagawa, S.; Noyori, R. *J. Org. Chem.* **1986**, *51*, 629-632.
9. Toriumi, K.; Ito, T.; Takaya, H.; Souchi, T.; Noyori, R. *Acta Crystallogr., Sect B* **1982**, *B38*, 807-812.
10. (a) Tani, K.; Yamagata, T.; Akutagawa, S.; Kumobayashi, H.; Taketomi, T.; Takaya, H.; Miyashita, A.; Noyori, R.; Otsuka, S. *J. Am. Chem. Soc.* **1984**, *106*, 5208-5217; (b) Tani, K.; Yamagata, T.; Otsuka, S.; Kumobayashi, H.; Akutagawa, S.; *Org. Synth.* **1988**, *67*, 33.
11. Ohta, T.; Takaya, H.; Noyori, R. *Inorg. Chem.* in press.
12. Ohta, T.; Takaya, H.; M. Kitamura, M.; Nagai, K.; Noyori, R. *J. Org. Chem.* **1987**, *52*, 3174-3176.
13. (a) Noyori, R.; Ohta, M.; Hsiao, Y.; Kitamura, M.; Ohta, T.; Takaya, H. *J. Am. Chem. Soc.* **1986**, *108*, 7117-7119; (b) Kitamura, M.; Hsiao, Y.; Noyori, R.; Takaya, H. *Tetrahedron Lett.* **1987**, *28*, 4829-4832.
14. (a) Takaya, H.; Ohta, T.; Sayo, N.; Kumobayashi, H.; Akutagawa, S.; Inoue, S.; Kasahara, I.; Noyori, R. *J. Am. Chem. Soc.* **1987**, *109*, 1596-1597, 4129; (b) Kitamura, M.; Kasahara, I.; Manabe, K.; Noyori, R.; Takaya, H. *J. Org. Chem.* **1988**, *53*, in press.
15. Noyori, R.; Ohkuma, T.; Kitamura, M.; Takaya, H.; Sayo, N.; Kumobayashi, H.; Akutagawa, S. *J. Am. Chem. Soc.* **1987**, *109*, 5856-5858.
16. Kitamura, M.; Ohkuma, T.; Inoue, S.; Sayo, N.; Kumobayashi, H.; Akutagawa, S.; Ohta, T.; Takaya, H.; Noyori, R. *J. Am. Chem. Soc.* **1988**, *110*, in press.

Appendix

Chemical Abstracts Nomenclature (Collective Index Number); (Registry Number)

(R)-(+)- and (S)-(-)-2,2'-Bis(diphenylphosphino)-1,1'-binaphthyl (BINAP):
Phosphine, [1,1'-binaphthalene]-2,2'-diylbis-[diphenyl-, (R)- or (S)- (10);
[(R)-: 76189-55-4; (S)-: 76189-56-5]

2,2'-Dibromo-1,1'-binaphthyl: 1,1'-Binaphthalene, 2,2'-dibromo- (10);
(74866-28-7)

1,1'-Bi-2-naphthol: [1,1'-Binaphthalene]-2,2'-diol (8); (602-09-5); [1,1'-Binaphthalene]-2,2'-diol, (±)- (9); (41024-90-2)

(±)-2,2'-Bis(diphenylphosphinyl)-1,1-binaphthyl [(±)-BINAPO]:
Phosphine oxide, [1,1'-binaphthalene]-2,2'-diylbis[diphenyl-,
(±)- (11); (86632-33-9)

Diphenylphosphinyl chloride: Phosphinic chloride, diphenyl-
(8,9); (1499-21-4)

(-)-2,3-O-Dibenzoyl-L-tartaric acid monohydrate [(-)-DBT monohydrate]:
Butanedioic acid, 2,3-bis(benzyloxy)-, [R-(R*,R*)]- (9); (2743-38-6)

(S)-(-)-2,2'-Bis(diphenylphosphinyl)-1,1'-binaphthyl [(S)-(-)-BINAPO]:
Phosphine oxide, [1,1'-binaphthalene]-2,2'-diylbis[diphenyl-, (S)- (11);
(94041-18-6)

(+)-2,3-O-Dibenzoyl-D-tartaric acid monohydrate [(+)-DBT monohydrate]:
Tartaric acid, dibenzoate, (-)- (8); Butanedioic acid, 2,3-bis(benzoyloxy)-,
[S-(R*,R*)]- (9); (17026-42-5)

(R)-(+)-2,2'-Bis(diphenylphosphinyl)-1,1'-binaphthyl [(R)-(+)-BINAPO]: Phosphine oxide, [1,1'-binaphthalene]-2,2'-diylbis[diphenyl-, (R)- (11); (94041-16-4)

Trichlorosilane: Silane, trichloro- (8,9); (10025-78-2)

(R)-(−)-N,N-DIETHYL-(E)-CITRONELLALENAMINE AND (R)-(+)-CITRONELLAL VIA ISOMERIZATION OF N,N-DIETHYLGERANYLAMINE OR N,N-DIETHYLNERYLAMINE

(1-(E), 6-Octadienylamine, (R)-(−)-N,N-diethyl-3,7-dimethyl-
and 6-octenal, (R)-(+)-3,7-dimethyl-)

A. $[RhCl(1,5-C_8H_{12})]_2$ + 2 (+)-BINAP $\xrightarrow{\text{AgClO}_4}{\text{acetone}}$ $[Rh\{(+)\text{-BINAP}\}(1,5-C_8H_{12})]\,ClO_4$

$[RhCl(1,5-C_8H_{12})]_2$ + 2 (−)-BINAP $\xrightarrow{\text{AgClO}_4}{\text{acetone}}$ $[Rh\{(−)\text{-BINAP}\}(1,5-C_8H_{12})]\,ClO_4$

B. [geranylamine NEt$_2$] $\xrightarrow{[Rh\{(−)\text{-BINAP}\}(1,5-C_8H_{12})]\,ClO_4}{\text{THF}}$ (−) [citronellal enamine NEt$_2$]

[nerylamine NEt$_2$] $\xrightarrow{[Rh\{(+)\text{-BINAP}\}(1,5-C_8H_{12})]\,ClO_4}{\text{THF}}$ (−) [citronellal enamine NEt$_2$]

C. (−) [citronellal enamine NEt$_2$] $\xrightarrow{CH_3CO_2H}$ (+) [citronellal CHO]

Submitted by Kazuhide Tani,[1a] Tsuneaki Yamagata,[1a] Sei Otsuka,[1a] Hidenori Kumobayashi,[1b] and Susumu Akutagawa.[1b]
Checked by David Coffen, Louis A. Portland, Bryant Rossiter, and Gabriel Saucy.

1. Procedure

Caution! All manipulations for the preparation of the transition metal complexes and the catalytic isomerization should be carried out under dry nitrogen or argon. All solvents used are distilled under dry nitrogen over metallic sodium, or after drying over calcium sulfate (in the case of acetone) immediately before use.

A. [(+)-2,2'-Bis(diphenylphosphino)-1,1'-binaphthyl]-(η^4-1,5-cyclooctadiene)rhodium(I) perchlorate. [Rh{(+)-BINAP}(1,5-C_8H_{12})] ClO_4. A dry, 250-mL Schlenk flask, filled with dry nitrogen or argon and equipped with a magnetic stirring bar, is charged with 0.53 g (1.08 mmol) of chloro-(1,5-cyclooctadiene)rhodium(I) dimer (Note 1). Then 40 mL of dry acetone is added using an air-tight syringe. The flask is protected from light by wrapping it with aluminum foil and 0.45 g (2.17 mmol) of silver perchlorate is added to the stirred suspension. The mixture is stirred for 1 hr at ambient temperature. The colorless precipitate of silver chloride is removed by suction filtration under argon through a stainless steel cannula fitted with a filter tip (Note 2). The precipitates are washed with 5 mL of acetone. To the pale orange filtrate and the washings is added 1.34 g (2.16 mmol) of (+)-2,2'-bis(diphenylphosphino)-1,1'-binaphthyl ((+)-BINAP) (Note 3) and the resulting dark red solution is stirred under argon for 18 hr at ambient temperature. The reaction mixture is concentrated under reduced pressure (60 mm) to ca. 3 mL. Then 30 mL of ether is slowly added with a syringe. The resulting mixture is stirred at ambient temperature for 18 hr. The orange solid is filtered off under argon and washed with 5 mL of ether. The crude product is dissolved in 50 mL of dry acetone and the solution is concentrated to ca. 3 mL under 60 mm pressure. Dry ether (30 mL) is slowly added and the mixture is stirred for 18 hr. The deep orange crystals are collected by filtration, washed with 5 mL of dry ether, and dried under vacuum to afford 2.07 g of recrystallized product, 230-235°C (dec) (Note 4).

[(-)-2,2'-Bis(diphenylphosphino)-1,1'-binaphthyl](η^4-1,5-cyclooctadiene)-rhodium(I) perchlorate: [Rh{(-)-BINAP}(1,5-C_8H_{12})] ClO_4 is similarly prepared from 0.53 g (1.08 mmol) of chloro-(1,5-cyclooctadiene)rhodium(I) dimer, 0.45 g (2.17 mmol) of silver perchlorate, and 1.34 g (2.16 mol) of (-)-2,2'-bis(diphenylphosphino)-1,1'-binaphthyl ((-)-BINAP) (Note 5) in acetone. The crude product is recrystallized from acetone-ether as above to give 2.02 g of orange powder, mp 230°C (dec) (Note 6).

B. *(R)-(-)-N,N-Diethyl-(E)-citronellalenamine [(R)-(-)-N,N-diethyl-3,7-dimethyl-1(E),6-octadienylamine]. From N,N-diethylgeranylamine.* A 500-mL, three-necked, round-bottomed flask equipped with a magnetic stirrer, a reflux condenser and an argon inlet is charged with 373 mg (0.40 mmol) of [(-)-2,2'-bis(diphenylphosphino)-1,1'-binaphthyl](η^4-1,5-cyclooctadiene)rhodium(I) perchlorate. The flask is evacuated and refilled with argon four times. Another 500-mL, three-necked, round-bottomed flask is charged with 83.8 g (0.40 mol) of N,N-diethylgeranylamine (Note 7) and 250 mL of distilled tetrahydrofuran (Note 8) under an argon blanket. This solution is transferred under argon by cannula to the flask containing the catalyst, which is then evacuated and refilled with argon twice. The reaction mixture is stirred and heated at reflux for 21 hr (Note 9). The solution is cooled to room temperature and the solvent is removed under vacuum (60 mm) at 45°C. The residue is vacuum distilled through a 10-cm Vigreux column to give 78.7 g (93.9%) of (R)-(-)-N,N-diethyl-(E)-citronellalenamine as a colorless liquid, bp 84-85°C (1.1 mm), $[\alpha]_D^{25}$ -66.5° (hexane, *c* 10.2). The product is 97.2% chemically pure by GLC analysis (Notes 10, 11 and 12).

From N,N-diethylnerylamine. Similarly, 83.8 g (0.40 mol) of N,N-diethylnerylamine (Note 13) is isomerized with 373 mg (0.40 mmol) of [(+)-2,2'-bis(diphenylphosphino)-1,1'-binaphthyl](η^4-1,5-cyclooctadiene)rhodium(I) perchlorate catalyst in 250 mL of THF (Note 8) at reflux for 70 hr (Note 9) to give 77.2 g (92.1%) of (R)-(-)-N,N-diethyl-(E)-citronellalenamine, $[\alpha]_D^{25}$ -66.9° (hexane, *c* 10.7). The chemical purity of the product is 91.7% by GLC (Notes 10, 12, and 14).

C. *(R)-(+)-Citronellal. [(R)-(+)-3,7-Dimethyl-6-octanal].* A 500-mL, round-bottomed flask equipped with a magnetic stirrer and an ice bath is charged with 33.4 g (0.16 mol) of (R)-(-)-N,N-diethyl-(E)-citronellalenamine in 80 mL of ether at 0°C. To this stirred solution is added 80 mL of a 1:4 glacial acetic acid-deionized water solution in one portion (Note 15). The reaction mixture is stirred for 5 min at 0°C and then at room temperature for 25 min. The ether layer is separated and washed successively with 50 mL of water, two 50-mL portions of saturated aqueous sodium bicarbonate solution, 50 mL of water, and 50 mL of saturated brine. The ether layer is dried over anhydrous sodium sulfate and filtered (Note 16). The ether solution is concentrated at 40°C under reduced pressure (60 mm) to a pale yellow liquid. The liquid is distilled through a 10-cm Vigreux column to give 22.5 g (91.4%) of (+)-citronellal as a colorless liquid, bp 79-80°C (7 mm), $[\alpha]_D^{25}$ +15.7° (neat, d 0.851). The product is 99.4% chemically pure by GLC and has an optical purity of 95.2% (Note 17).

2. Notes

1. Chloro(η^4-1,5-cyclooctadiene)rhodium(I) dimer, [RhCl(1,5-C_8H_{12})] can be prepared according to the method described in *Inorganic Syntheses*.[2] The checkers used material purchased from the Aldrich Chemical Company, Inc.

2. This can also be done as previously described in *Inorganic Syntheses*.[3]

3. (R)-(+)-2,2'-Bis(diphenylphosphino)-1,1'-binaphthyl was prepared according to Takaya, H.; Akutagawa, S.; Noyori, R. *Org. Synth.* **1988**, *67*, 20.

4. The submitters obtained 1.5-1.7 g of analytically pure material with mp 164°C (dec). The ^{1}H NMR (CD_2Cl_2) of this product showed signals at δ: 2.00-2.62 (m, 8 H, -C$\underline{H}_2$-), 4.58 (br signal, 2 H, -C$\underline{H}$=), 4.84 (br signal, 2 H, -C$\underline{H}$=), 6.42-8.22 ppm (m, 32 H, arom.); Anal. Calcd for ($C_{52}H_{44}ClO_4Rh$): C, 66.93; H, 4.75; Cl, 3.80. Found: C, 66.66; H, 4.89; Cl, 3.92.

5. (S)-(-)-2,2'-Bis(diphenylphosphino)-1,1'-binaphthyl was prepared according to the *Organic Syntheses* procedure; see Note 3.

6. The submitters used THF-ether for recrystallization, in which case the product was obtained as a THF-solvated complex, [Rh{(-)-BINAP}(1,5-C_8H_{12})]ClO_4 THF, in the form of deep orange crystals, mp 153°C (dec). It's ^{1}H NMR (CD_2Cl_2) spectrum showed clearly the presence of the solvating tetrahydrofuran; δ: 1.60-2.00 (m, 4 H, -C$\underline{H}_2$- of THF), 2.00-2.62 (m, 8 H, -C$\underline{H}_2$-), 3.40-3.80 (m, 4 H, -C$\underline{H}_2$O- of THF), 4.58 (br signal, 2 H, -C$\underline{H}$=), 4.82 (br, signal, 2 H, -C$\underline{H}$=), 6.40-7.97 (m, 32 H, arom.); Anal. Calcd for ($C_{52}H_{44}ClO_4RhC_4H_8O$): C, 66.90; H, 5.21; Cl, 3.53. Found: C, 66.55; H, 5.62; Cl, 3.68.

7. N,N-Diethylgeranylamine was prepared according to the method described in *Org. Synth.* **1988**, *67*, 44. The submitters used material of 99.7% purity (determined by GLC, Hitachi 063 with OV-101 in Fused Silica, 0.2 mm x 25 m column) that was obtained by careful distillation through a column with one hundred theoretical plates, bp 70°C (2 mm), and redistilled over calcium hydride before use. The checkers used a 32-cm Goodloe column and collected the amine at 105°C (2.4 mm), which had a purity of 100% (GLC).

8. Acetone or methanol also may be used. However, tetrahydrofuran gives the best results. The solvents should be strictly dry because water deactivates the catalyst.

9. The submitters ran the reaction at 60°C for 24 hr.

10. Purity of the product and progress of the isomerization can be determined by GLC (Triton X 305 packed in a 0.2 mm x 30 m glass column from Gasukuro Kogyo Co. Ltd., or a column-crosslinked 5% phenylmethyl silicone, 25 m high performance capillary column supplied by Hewlett Packard. Starting temperature, 125°C; rate 3°/min; final temperature, 180°C.) The product enamine is 100% (E)-isomer. The spectral properties of the product are as follows: IR (neat) cm^{-1}: 3045, 2960, 2915, 2860, 2720, 1660, 1655, 1450, 1377, 1298, 1245, 1196, 1100, 985, 937, 886, 830, 784, 740; ^{1}H NMR (acetone d$_6$) δ: 0.96 (d, 3 H, J = 7.2), 1.01 (t, 6 H, J = 7.1), 1.1-1.4 (m, 2 H), 1.58 (br s, 3 H), 1.66 (br s, 3 H), 1.8-2.1 (m, 3 H), 2.92 (q, 4 H, J = 7.1), 3.93 (d of d, 1 H, J = 14.6, 8.4), 5.11 (br t, 1 H, J = 7.3), 5.79 (d of d, 1 H, J = 14.6, 0.8).

11. The submitters report 92-96% yields of 98% chemically pure product, bp 80-82°C (1 mm), $[\alpha]_D^{25}$ -74.3° (hexane, *c* 10.0). The specific rotation was corrected for this purity (Note 12). A specific rotation of $[\alpha]_D^{21}$ -77.6° (hexane) is estimated for the pure enamine.

12. The product is very moisture sensitive and should be handled under dry nitrogen or argon. Despite this precaution, the product is always contaminated by small amounts of (+)-citronellal, which has an optical rotation opposite to that of the enamine. To determine the optical purity of the product enamine, the specific rotation measured therefore must be corrected for the (+)-citronellal impurity. It is more reliable to base optical purity on the specific rotation of the citronellal obtained by hydrolysis of the enamine (Part C). The absolute method using HPLC of the diastereomeric amide derivative[4] also may be useful as a check of the optical purity.

13. N,N-Diethylnerylamine was prepared according to the method described in *Org. Synth.* **1988**, *67*, 48, and distilled over calcium hydride and stored under nitrogen below -20°C. GLC-MS analysis (Hitachi 063 and Hitachi RMU-6MG with OV-101 in Fused Silica, 0.2 mm x 25 m column) of N,N-diethylnerylamine used by the submitters showed a purity of 94.9%, and contained N,N-diethyl-2-ethylidene-6-methyl-5-heptenylamine (0.2%), N,N-diethyl-2,7-dimethyl-2,6-octadienylamine (1.5%), N,N-diethyl-3-methylene-7-methyl-6-octenylamine (2.1%), and unidentifiable products (1.3%) as impurities, but not the (E)-isomer, N,N-diethylgeranylamine. The checkers distilled the nerylamine through a 32-cm Goodloe column, bp 102°C (3 mm), and achieved a purity of 98.6%.

14. The submitters obtained 90-95% yields of 94% chemically pure product, $[\alpha]_D^{24}$ -73.1° (hexane, *c* 10.0; corrected for the citronellal impurity).

15. The submitters conducted the hydrolysis step in a mixture of 2N sulfuric acid and toluene. The acid is added dropwise at 0°C at a rate that keeps the pH of the mixture at 4-5. Control of pH is critical.

16. Rapid work-up after the acid-hydrolysis is desirable.

17. A specific rotation of $[\alpha]_D^{25}$ +16.5° (neat) is reported for optically pure citronellal.[5]

3. Discussion

(R)-(+)-Citronellal is a useful, key intermediate for the preparation of several important, optically active compounds such as citronellol, 1-menthol,[5] muscone,[6] and α-tocopherol.[7] The optical purity of citronellal from natural sources is at most 77% ee, however. This new procedure gives (R)-(+)-citronellal of high optical purity (over 95% ee).

The intermediate enamine, (R)-(-)-N,N-diethyl-(E)-citronellalenamine, is also a key intermediate for preparation of useful, optically-active compounds, e.g., (+)-7-hydroxydihydrocitronellal.[8]

Isomerization of N,N-diethylnerylamine with [Rh{(-)-BINAP}(1,5-C_8H_{12})]ClO_4 and N,N-diethylgeranylamine with [Rh{(+)-BINAP}(1,5-C_8H_{12})] ClO_4 under similar conditions to those described in the text ([substrate]/[Rh] = 100, in THF, 40°C, 23 hr) proceed equally efficiently to give (+)-N,N-diethyl-(E)-citronellalenamine in good chemical (97% and 95%, respectively) and optical yields (92% and 96%, respectively). Thus, unnatural (S)-(-)-citronellal with high optical purity (over 95% ee) can also be prepared by the new procedure. The substrates must be geometrically pure and the chiral ligands enantiomerically pure in order to achieve optimal results.

If extreme care is taken to purify the substrate and the solvent, the [substrate]/[Rh] ratio can be raised much higher. For example, N,N-diethylgeranylamine can be isomerized in tetrahydrofuran in the presence of 0.00125 mol% of [Rh{(-)-BINAP}(1,5-C_8H_{12})] ClO_4 at 100°C in a glass autoclave during 3-7 hr to give (-)-N,N-diethyl-(E)-citronellalenamine with 97% ee in almost quantitative yield.

With the same catalyst systems, other prochiral N-alkyl- or N,N-dialkylallylamines can also be isomerized efficiently to the corresponding optically active imines or (E)-enamines, respectively. For example, with [Rh{(+)-BINAP}(1,5-C_8H_{12})] ClO_4, ([substrate]/[Rh] = 100, in THF, 40°C, 23 hr) a secondary allylamine, N-cyclohexylgeranylamine, and an allylamine with styrene-type conjugation, N,N-dimethyl-3-phenyl-2(E)-butenylamine, are isomerized to give (S)-(-)-N-cyclohexylcitronellalimine (chemical yield, 100%; optical yield, 96%) and (R)-(-)-N,N-dimethyl-3-phenyl-1(E)-butenylamine (chemical yield, 84%; optical yield, 90%), respectively.

Under similar conditions, various kinds of N,N-dialkylamines with substituents at β- or γ-positions, e.g., N,N-dimethyl-2-propenylamine, N,N-dimethyl-2(E)-butenylamine, N,N-dimethyl-2-methyl-2-propenylamine, and N,N-dimethyl-3-methyl-2-butenylamine can also be isomerized to the corresponding (E)-enamine in medium to good yields (60, 52, 97, and 100%). However, N,N-dimethyl-2(E)-butenylamine and N-phenyl- or N,N-diphenylgeranylamine were found to be poor substrates.

1. (a) Department of Chemistry, Faculty of Engineering Science, Osaka University, Toyonaka, Osaka, 560 Japan; (b) Central Research Laboratory, Takasago Perfumery Co., Ltd., 31-36, 5-chome, Kamata, Ohta-ku, Tokyo, 144 Japan.
2. Giordano, G.; Crabtree, R. H. *Inorg. Synth.* **1979**, *19*, 218.
3. Tatsuno, Y.; Yoshida, T.; Otsuka, S. *Inorg. Synth.* **1979**, *19*, 220.
4. Valentine, Jr., D.; Chan, K. K.; Scott, C. G.; Johnson, K. K.; Toth, K.; Saucy, G. *J. Org. Chem.* **1976**, *41*, 62.
5. Sully, B. D.; Williams, P. L. *Perfum. Essent. Oil Rec.* **1968**, *59*, 365; *Chem. Abstr.* **1968**, *69*, 38703u.
6. Utimoto, K.; Tanaka, M.; Kitai, M.; Nozaki, H. *Tetrahedron Lett.* **1978**, 2301.
7. Chan, K.-K.; Cohen, N.; De Noble, J. P.; Specian, Jr., A. C.; Saucy, G. *J. Org. Chem.* **1976**, *41*, 3497.
8. Ishino, R.; Kumanotani, J. *J. Org. Chem.* **1974**, *39*, 108.

Appendix
Chemical Abstracts Nomenclature (Collective Index Number); (Registry Number)

(R)-(-)-N,N-Diethyl-(E)-citronellalenamine: 1,6-Octadien-1-amine, N,N-diethyl-3,7-dimethyl-, [R-(E)]- (10); (67392-56-7)

(R)-(+)-Citronellal: 6-Octenal, 3,7-dimethyl-, (R)-(+)- (8,9); (2385-77-5)

N,N-Diethylgeranylamine: 2,6-Octadien-1-amine, N,N-diethyl-3,7-dimethyl-, (E)- (9); (40267-53-6)

N,N-Diethylnerylamine: 2,6-Octadien-1-amine, N,N-diethyl-3,7-dimethyl-, (Z)- (9); (40137-00-6)

[(+)-2,2'-Bis(diphenylphosphino)-1,1'-binaphthyl-(η^4-1,5-cyclooctadiene)rhodium(I) perchlorate: Rhodium(1+), [[1,1'-binaphthalene]-2,2'-diylbis[diphenylphosphine]-P,P'][(1,2,5,6-η)-1,5-cyclooctadiene]-, stereoisomer, perchlorate (11); (82822-45-5)

Di-μ-chlorobis(η^4-1,5-cyclooctadiene)dirhodium(I): Rhodium, di-μ-chlorobis(1,5-cyclooctadiene)di- (8); Rhodium, di-μ-chlorobis[(1,2,5,6-η)-1,5-cyclooctadiene]di- (9); (12092-47-6) silver perchlorate: Perchloric acid, silver(1+) salt, monohydrate (8,9); (14242-05-8)

(R)-(+)- and (S)-(-)-2,2'-Bis(diphenylphosphino)-1,1'-binaphthyl [BINAP]: Phosphine, [1,1'-binaphthalene]-2,2'-diylbis[diphenyl-, (R)- or (S)- (10); [(R)-: (76189-55-4); (S)-: (76189-56-5)]

[(-)-2,2'-Bis(diphenylphosphino)-1,1'-binaphthyl](η^4-1,5-cyclooctadiene)rhodium(I) perchlorate: Rhodium(1+), [[1,1'-binaphthalene]-2,2'-diylbis[diphenylphosphine]-P,P'][(1,2,5,6-η)-1,5-cyclooctadiene]-, stereoisomer, perchlorate (11); (82889-98-3)

ADDITION OF DIALKYLAMINES TO MYRCENE: N,N-DIETHYLGERANYLAMINE

(2,6-Octadien-1-amine, N,N-diethyl-3,7-dimethyl-,(E))

[structure: myrcene] $\xrightarrow{\text{(C}_2\text{H}_5)_2\text{NH}}_{\text{Li or BuLi}}$ [structure: N,N-diethylgeranylamine with N(C$_2$H$_5$)$_2$ group]

Submitted by Kunihiko Takabe,[1] Takashi Katagiri,[1] Juntaro Tanaka,[1], Tsutomu Fujita,[2] Shoji Watanabe,[2] and Kyoichi Suga.[2]
Checked by Alan J. Chalk, Laszlo V. Wertheimer, and Gabriel Saucy.

1. Procedure

In a 50-mL, round-bottomed glass reactor equipped with a magnetic stirring bar are placed 13.60 g (74 mmol) of myrcene (Note 1), 10.29 g (141 mmol) of diethylamine (Note 2) and 0.185 g (0.0267 g-atom) of metallic lithium cut into small pieces. The vessel is flushed with dry nitrogen, and is sealed. The solution is heated to 55°C in a water-bath, and is stirred for 5 hr. The vessel is cooled to room temperature and the contents are poured into 30 mL of ice-water. The upper organic layer is separated, and the aqueous layer is extracted with 20-mL portions of diethyl ether. The combined organic layer is washed with aqueous sodium sulfate solution, dried over anhydrous sodium sulfate, and evaporated under reduced pressure. Distillation (Note 3) of the residual liquid affords 1.2-2.0 g of unreacted myrcene and 12.66-13.28 g (74-77%) of the product as a colorless liquid, bp 67-68°C (0.5 mm). GLC analysis indicated that the product contained 91.2-92.5% of N,N-diethylgeranylamine and other isomers (Note 4).

The checkers found that the use of benzene (20 mL) as a solvent increased the selectivity for N,N-diethylgeranylamine to 94% (Notes 5, 6).

A similar reaction of myrcene with dipropylamine (50°C, 20 hr) afforded *N,N-dipropylgeranylamine* (80%; bp 93-94°C, 1 mm).

2. Notes

1. Myrcene, obtained from SCM Organic Chemicals (also available from Takasago Perfumery Company, Ltd., in Japan), was distilled (bp 69-70°C, 20 mm) prior to use. GLC analysis (Triton X-305, 0.28 mm x 30 m, 80-120°C) showed that the fraction contained 74% myrcene. The submitters used 80% pure myrcene.

2. Diethylamine, obtained from Aldrich Chemical Company, Inc. (also available from Nakarai Chemicals, Ltd., in Japan), was distilled from calcium hydride before use.

3. A short-path distillation apparatus was used in order to prevent loss of the product.

4. GLC analysis (Triton X-305; 0.28 mm x 30 m, 80-160°C) showed a composition of 92% N,N-diethylgeranylamine and other isomers which were identified by the checkers.[3] The spectral properties of N,N-diethylgeranylamine are as follows: IR (neat) cm^{-1}: 1660, 1200, 1165, 1050, 830; ^{1}H-NMR (CDCl$_3$, 60 MHz) δ: 0.96 (t, 6 H, J = 7), 1.44-1.67 (m, 6 H), 1.85-2.15 (m, 4 H), 2.40 (q, 4 H, J = 7), 2.92 (d, 2 H, J = 6.5), 4.77-5.30 (m, 2 H).

5. The submitters preferred butyllithium in hexane in place of lithium metal, as follows: In a 50-mL, round-bottomed flask equipped with a magnetic stirring bar, are placed 4.08 g (24 mmol) of myrcene and 3.29 g (45 mmol) of diethylamine under nitrogen. The mixture is cooled to 0°C using an ice bath,

and 3.0 mL (4.8 mmol) of a 1.60 M solution of butyllithium in hexane is added dropwise by syringe while stirring for 15 min. The flask is sealed, the solution is warmed to 50°C and stirred for 4 hr. The vessel is cooled to room temperature, and the contents are poured into 20 mL of cold water. The vessel is washed with 30 mL of diethyl ether and 10 mL of water. The upper layer is separated, and the aqueous layer is extracted twice with 20 mL of diethyl ether. The combined organic layers are washed with brine, dried over anhydrous sodium sulfate, and evaporated under reduced pressure. Distillation (Note 3) of the residual liquid affords a 0.31-0.42 g forerun of unreacted myrcene and 4.06-4.37 g of the product (77-83%) as a colorless liquid, bp 84-86°C (1.5); composition by GLC: 95.3% N,N-diethylgeranylamine and 4.5% other isomers.

6. The submitters report that sodium naphthalenide can be used in place of butyllithium. Using tetrahydrofuran as a solvent, they obtained N,N-diethylgeranylamine in 56% yield.

3. Discussion

N,N-Dialkylgeranylamines have been prepared by the reaction of dialkylamines with geranyl halides.[4,5] The procedure described here is a modification of one we reported earlier.[4,6] It is a simple, one-step synthesis of N,N-dialkylgeranylamines from myrcene and dialkylamines which are readily available bulk chemicals. The reaction proceeds stereoselectively, and yields are high.

N,N-Diethylgeranylamine is a key intermediate for the synthesis of industrially important acyclic monoterpenes such as geranyl acetate,[7] linalool,[8] citral[9] and citronellal.[10]

1. Department of Synthetic Chemistry, Faculty of Engineering, Shizuoka University, Johoku, Hamamatsu 432, Japan.
2. Department of Applied Chemistry, Faculty of Engineering, Chiba University, Yayoicho, Chiba 280, Japan.
3. Chalk, A. J.; Magennis, S. A. *Ann. N.Y. Acad. Sci.* **1980**, *333*, 286-301.
4. Takabe, K.; Katagiri, T.; Tanaka, J. *Bull. Chem. Soc., Jap.* **1973**, *46*, 222-225.
5. Rautenstrauch, V. *Helv. Chim. Acta* **1973**, *56*, 2492-2508.
6. Fujita, T.; Suga, K.; Watanabe, S. *Chem. Ind. (London)* **1973**, 231-232.
7. Fujita, T.; Suga, K.; Watanabe, S. *Aust. J. Chem.* **1974**, *27*, 531-535.
8. Takabe, K.; Katagiri, T.; Tanaka, J. *Tetrahedron Lett.* **1975**, 3005-3006.
9. Takabe, K.; Yamada, T.; Katagiri, T. *Chemistry Lett.* **1982**, 1987-1988.
10. Tani, K.; Yamagata, T.; Otsuka, S.; Akutagawa, S.; Kumobayashi, H.; Taketomi, T.; Takaya, H.; Miyashita, A.; Noyori, R. *J. Chem. Soc., Chem. Commun.* **1982**, 600-601.

Appendix
Chemical Abstracts Nomenclature (Collective Index Number); (Registry Number)

Myrcene: 1,6-Octadiene, 7-methyl-3-methylene- (8,9); (123-35-3)
N,N-Diethylgeranylamine: 2,6-Octadien-1-amine, N,N-diethyl-3,7-dimethyl-, (E)- (9); (40267-53-6)

TELOMERIZATION OF ISOPRENE WITH DIALKYLAMINE: N,N-DIETHYLNERYLAMINE

(2,6-Octadien-1-amine, N,N-diethyl-3,7-dimethyl-, (Z)-)

$$\text{isoprene} \xrightarrow[\text{BuLi or Li}]{(C_2H_5)_2NH} \text{N,N-diethylnerylamine–N}(C_2H_5)_2$$

Submitted by Kunihiko Takabe, Takashi Yamada, Takao Katagiri, and Juntaro Tanaka.[1]
Checked by Alan J. Chalk, Laszlo V. Wertheimer, and Gabriel Saucy.

1. Procedure

A 100-mL glass autoclave (Note 1) is charged with 34 g (0.50 mol) of isoprene (Note 2) and 7.3 g (0.10 mol) of diethylamine (Note 3) under nitrogen (Note 4). The contents are cooled to 3-5°C using an ice-salt bath, and 3.7 mL (0.006 mol) of a 1.62 M solution of butyllithium in hexane (Note 5) is added dropwise by syringe while stirring. The resulting slightly yellow solution is blanketed carefully with dry nitrogen, and the vessel is sealed. The solution is heated to 65°C, and stirred for 13 hr. The vessel is then cooled using an ice-salt bath, the contents are poured into 15 mL of water, and the organic layer is separated. The aqueous phase is extracted with 30 mL of diethyl ether. The organic layers are combined, dried over anhydrous sodium sulfate, filtered, and concentrated under reduced pressure with a rotary evaporator. The residual liquid is distilled under reduced pressure (Note 6) to afford 16.1-18.0 g (77-86%) of N,N-diethylnerylamine as a colorless liquid, bp 83-84°C (1.5 mm) (Notes 7 and 8).

2. Notes

1. The glass autoclave must be used with appropriate precaution. A pressure of about 20 p.s.i. is generated, so a low positive pressure reactor, such as a shielded Fischer-Porter Bottle (Fischer and Porter Company, Lab-Coest Division, Warminster, PA 18974), can be used instead. Alternatively the reaction may be carried out without pressure equipment in 3 days.[2]

2. Isoprene, obtained from Aldrich Chemical Company, Inc. (also available from Nakarai Chemicals, Ltd. in Japan), was dried over Linde-type 4Å molecular sieves for at least 1 day and freshly distilled prior to use.

3. Diethylamine, obtained from Aldrich Chemical Company, Inc., was distilled from calcium hydride before use.

4. A slow stream of dry nitrogen was passed through an inverted funnel that was placed over the vessel.

5. Butyllithium in hexane was obtained from Aldrich Chemical Company, Inc. (also available from Nakarai Chemicals, Ltd. in Japan), and titrated before use.[3] Lithium (0.055 g, 0.0079 g-atom), cut into small pieces (ca. 2 mm), can be used in place of the butyllithium-hexane solution.

6. A forerun (1.4-0.6 g) boiling at 52-56°C (16 mm) consists mainly of N,N-diethyl-2-methyl-2-butenylamine and N,N-diethyl-3-methyl-2-butenylamine.

7. GLC analysis (7% Apiezon L on 60-80 mesh Gaschrom Q, 3 mm x 3 m, 170°C) showed that the product is over 99% isomerically pure. The spectral properties of N,N-diethylnerylamine are as follows: IR (neat) cm^{-1}: 1660, 1200, 1165, 1050, 830; ^{1}H NMR (CDCl$_3$) δ: 1.03 (t, 6 H, J = 7, CH$_3$-), 1.45-1.7 (m, 6 H, CH$_3$-C=), 2.02-2.2 (m, 4 H, -CH$_2$-C=), 2.50 (q, 4 H, J = 7, -N-CH$_2$-), 3.05 (d, 2 H, J = 6.5, -N-CH$_2$-C=), 4.9-5.5 (m, 2 H, -CH=).

8. The submitters report that similar reactions of isoprene with dimethylamine (55°C, 10 hr) and dipropylamine (65°C, 15 hr) afforded N,N-dimethylnerylamine (69%), bp 85-86°C (9 mm) and N,N-dipropylnerylamine (77%), bp 86-87°C (1 mm), respectively.

3. Discussion

The procedure described here is essentially that reported earlier[4] and modified by subsequent experience.[5] It is a simple, general method for the synthesis of an N,N-dialkylnerylamine from isoprene and a dialkylamine. N,N-Dialkylnerylamines can also be prepared by the reaction of a dialkylamine with neryl chloride[6] by a modification of Sandler's method.[7] However, pure nerol, the starting material of neryl chloride, is expensive, and not easily available commercially. The present method illustrates a mild and convenient one-step reaction for the preparation of an N,N-dialkylnerylamine. In addition, the starting materials are readily accessible, the reaction proceeds stereoselectively, and the yields of the product are generally high. This process consists of the initial addition of lithium dialkylamide to isoprene, followed by the propagation of the resulting intermediate, and the termination by dialkylamine.

N,N-Dialkylnerylamines serve as convenient precursors for naturally occurring acyclic monoterpenes such as linalool,[8] citronellal,[9] and citral.[10]

1. Department of Synthetic Chemistry, Faculty of Engineering, Shizuoka University, Johoku, Hamamatsu 432 Japan.
2. Chalk, A. J.; Magennis, S. A.; Wertheimer, L. V. In "Fundamental Research In Organometallic Chemistry", Tsutsui, M.; Ishii, Y.; Yaozeng, H., Eds.; Van Nostrand Reinhold: New York, 1982; p. 851.
3. Kofron, W. G.; Baclawski, L. M. *J. Org. Chem.* **1976**, *41*, 1879-1880.
4. Takabe, K.; Katagiri, T.; Tanaka, J. *Tetrahedron Lett.* **1972**, 4009-4012.
5. Takabe, K.; Katagiri, T.; Tanaka, J. *Chemistry Lett.* **1977**, 1025-1026.
6. Stork, G.; Grieco, P. A.; Gregson, M. *Org. Synth., Collect. Vol. VI* **1988**, 638.
7. Sandler, S. R.; Karo, W. "Organic Functional Group Preparations"; Academic Press: New York, 1968; Vol. 1, pp 324-325.
8. Takabe, K.; Katagiri, T.; Tanaka, J. *Tetrahedron Lett.* **1975**, 3005-3006.
9. Tani, K.; Yamagata, T.; Otsuka, S.; Akutagawa, S.; Kumobayashi, H.; Taketomi, T.; Takaya, H.; Miyashita, A.; Noyori, R. *J. Chem. Soc., Chem. Commun.* **1982**, 600-601.
10. Takabe, K.; Yamada, T.; Katagiri, T. *Chemistry Lett.* **1982**, 1987-1988.

Appendix
Chemical Abstracts Nomenclature (Collective Index Number);
(Registry Number)

N,N-Diethylnerylamine: 2,6-Octadien-1-amine, N,N-diethyl-3,7-dimethyl-, (Z)- (9); (40137-00-6)

(S)-N,N-DIMETHYL-N'(1-tert-BUTOXY-3-METHYL-2-BUTYL)FORMAMIDINE

(Methanimidamide, N'-[1-[(1,1-dimethylethoxy)methyl]-2-methylpropyl]-N,N-dimethyl-, (S)-)

Submitted by Daniel A. Dickman, Michael Boes, and Albert I. Meyers.[1]
Checked by Jeffrey Romine and Leo A. Paquette.

1. Procedure

A. *(S)-N-Formyl-O-tert-butylvalinol* (**1**). In a 100-mL, round-bottomed flask, 20.6 g (200 mmol) of (S)-valinol (Note 1) and 16 g (216 mmol) of ethyl formate (Note 2) are heated at reflux under a nitrogen atmosphere for 1 hr. Excess ethyl formate is removed under reduced pressure and the oil is triturated with dry ether until a yellow solid appears. This material is dissolved in 260 mL of dry dioxane (Note 3) in a 1000-mL pressure bottle (Note 4) which is equipped with a magnetic stirring bar and is immersed in an ice-water bath. The bottle is immediately charged with ca. 260 mL of liquid isobutene (Note 5) and 75 mL of boron trifluoride etherate is rapidly added.

The pressure bottle is sealed with a stopper, removed from the ice bath, and stirred at room temperature for 3 hr (Note 6). In a fume hood, excess isobutene is removed from the resulting clear solution by carefully cracking the seal of the stopper. When the gas ceases to discharge, the stopper is removed and the solution is poured into a 1000-mL separatory funnel containing 250 mL of 2 N sodium hydroxide and is extracted twice with 100 mL of dichloromethane. The organic layer is washed with 100 mL of brine and dried over anhydrous magnesium sulfate. The organic solvent is removed and the residue is distilled (Kugelrohr tube, 0.05 mm, 80-85°C bath temperature) to give 27-36 g (75-95%) of N-formyl-O-tert-butylvalinol (**1**) as a clear oil (Note 7).

B. *(S)-N,N-Dimethyl-N'-(1-tert-butoxy-3-methyl-2-butyl)formamidine* (**2**). In a 500-mL, round-bottomed flask 26 g (140 mmol) of the formamide from Part A is dissolved in 100 mL of ethanol and 200 mL of a 50% aqueous potassium hydroxide solution is added. The mixture is heated at reflux overnight; upon cooling, the reaction separates into colorless aqueous and organic layers. The two layers are extracted three times with 100 mL of ether and the combined organic layers are washed with 100 mL of brine. After the solution is dried over anhydrous potassium carbonate and filtered, the ether and ethanol are carefully removed under aspirator vacuum at ambient temperature. The crude amine is treated with 25 g (210 mmol) of N,N-dimethylformamide dimethyl acetal (Note 8) and the reaction mixture is heated under argon at 40°C for 1 hr. The solution is concentrated under reduced pressure and the crude product is distilled bulb-to-bulb (0.05 mm, 55-65°C) to give 25.7-27 g (86-91.5%) of (S)-N,N-dimethyl-N'-(1-tert-butoxy-3-methyl-2-butyl)formamidine (**2**) as a colorless liquid (Note 9).

2. Notes

1. (L)- or (S)-Valinol was purchased from Aldrich Chemical Company, Inc. and used without further purification. The preparation of (L)-valinol has been described: Smith, G. A.; Gawley, R. E. *Org. Synth.* **1985**, *63*, 136.

2. Ethyl formate was purchased from J. T. Baker Chemical Company.

3. Dioxane was distilled from lithium aluminum hydride.

4. A Kimble bottle (#15096) purchased from VWR Scientific (Cat. no. 16267-101) was employed.

5. Isobutene was purchased from Matheson Gas Products.

6. The two-layer system became a clear solution within 15 min.

7. The physical properties are as follows: IR (neat) cm^{-1}: 3300, 1660; ^{1}H NMR (CDCl$_3$, 300 MHz) δ: 0.79-0.88 (m, 6 H), 1.07 (s, 9 H), 1.82 (m, 1 H), 3.3 (m, 2 H), 3.75 (m, 1 H), 7.94 (d, 1 H, J = 12), 8.13 (d, 1 H, J = 1); $[\alpha]_D^{25}$ -59.6° (EtOH, *c* 3.5).

8. N,N-Dimethylformamide dimethyl acetal was purchased from Aldrich Chemical Company, Inc.

9. The physical properties are as follows: IR (neat) cm^{-1}: 1660; ^{1}H NMR (CDCl$_3$, 300 MHz) δ: 0.77 (d, 3 H, J = 6.5), 0.79 (d, 3 H, J = 6.5), 1.07 (s, 9 H), 1.72 (m, 1 H), 2.6-3.5 (m, 3 H), 2.73 (s, 6 H), 7.14 (s, 1 H); $[\alpha]_D^{25}$ -15.9° (THF, *c* 0.98).

3. Discussion

This procedure for the synthesis of N,N-dimethyl-N'-alkylformamidines is representative for both chiral and achiral alkyl groups. These compounds are used to activate a wide range of secondary amines toward metalation and alkylation and may be removed to furnish the α-alkylated amines.[2]

N,N-Dimethyl-N'-alkylformamidines may be prepared from dimethylformamide dimethyl acetal and a primary amine by heating for 1-5 hr[3] (eq. 1, Table I).

$$RNH_2 + Me_2NCH(OMe)_2 \xrightarrow{\Delta} \underset{R}{\underset{|}{N}}{-}CH{=}NMe_2 \quad (1)$$

It is also possible to simply heat "DMF-acetal" with an amino alcohol (e.g. valinol, leucinol, etc.) and obtain the hydroxy formamidine which can be directly silylated with Et_3SiCl, Me_3SiCl, or $t\text{-}BuMe_2SiCl$ at 0°C in dichloromethane (eq. 2).[2a] If these N,N-dimethylformamidines are required, this procedure has the advantage of eliminating the sometimes troublesome cleavage of a silyl ether during reaction with DMF-acetal.

$$H_2N\text{-}CH(R)\text{-}CH_2OH + \text{DMF-acetal} \longrightarrow Me_2N\text{-}CH{=}N\text{-}CH(R)\text{-}CH_2OH \xrightarrow[0°C]{R_3SiCl, CH_2Cl_2} Me_2N\text{-}CH{=}N\text{-}CH(R)\text{-}CH_2OSiR_3 \quad (2)$$

In addition to the exchange reaction[4] described previously,[5] formamidines derived from secondary amines can be prepared by forming the N-formyl derivative, which is treated successively with boron trifluoride etherate and the appropriate primary amine[2a] (eq. 3, Table II). However, this method is not satisfactory if sensitive groups (e.g., Me_3Si) are present on the amine since they are cleaved by the Meerwein reagent.

The main advantages of using the tert-butyl ether of the valinol formamidine are its stability to reaction conditions used in the asymmetric alkylation of amines and its ready recovery from these reactions for further use.

TABLE I

PREPARATION OF N,N-DIMETHYL-N'-ALKYLFORMAMIDINES

RNH$_2$	% yield
H$_2$N–CH(CH(CH$_3$)$_2$)–CH$_2$OMe	95
H$_2$N–CH(CH(CH$_3$)$_2$)–CH$_2$OSiEt$_3$	98
H$_2$N–CH(CH(CH$_3$)CH$_2$CH$_3$)–CH$_2$OMe	98
Me$_3$SiO–CH(Ph)–CH(OSiMe$_3$)–NH$_2$	96
2,2-dimethyl-1,3-dioxane with Ph and NH$_2$ substituents	98

TABLE II

PREPARATION OF FORMAMIDINES VIA N-FORMYL DERIVATIVES

RNH$_2$	(%)

(product: 3,4-dihydroisoquinolin-2(1H)-yl-N=CH—N—R)

RNH$_2$	(%)
H$_2$N–C(Ph)(H)(Me) (α-methylbenzylamine)	98
H$_2$N–CH(CH$_2$Ph)(CH$_2$OMe)	94
pinanyl-CH$_2$NH$_2$	98
methylpinanyl-CH$_2$NH$_2$	97

1. Department of Chemistry, Colorado State University, Fort Collins, CO 80523.
2. (a) Meyers, A. I.; Fuentes, L. M.; Kubota, Y. *Tetrahedron* **1984**, *40*, 1361; (b) Meyers, A. I.; Boes, M.; Dickman, D. A. *Angew. Chem., Intern. Ed. Engl.* **1984**, *23*, 458.
3. Abdulla, R. F.; Brinkmeyer, R. S. *Tetrahedron* **1979**, *35*, 1675.
4. Bredereck, H.; Effenberger, F.; Hofmann, A. *Chem. Ber.* **1964**, *97*, 61.
5. Meyers, A. I.; Boes, M., Dickman, D. A. *Org. Synth.* **1988**, *67*, 60.

Appendix
Chemical Abstracts Nomenclature (Collective Index Number); (Registry Number)

(S)-N,N-Dimethyl-N'-(1-tert-butoxy-3-methyl-2-butyl)formamidine: Methanimidamide, N'-[1-[(1,1-dimethylethoxy)methyl]-2-methylpropyl]-N,N-dimethyl-, (S)- (11); (90482-06-7)

(S)-N-Formyl-O-tert-butylvalinol: Formamide, N-[1-[(1,1-dimethylethoxy)methyl]-2-methylpropyl]-, (S)- (11); (90482-04-5)

Valinol: 1-Butanol, 2-amino-3-methyl-, L (8); 1-Butanol, 2-amino-3-methyl-, (S)- (9); (2026-48-4)

N,N-Dimethylformamide dimethyl acetal: Trimethylamine, 1,1-dimethoxy- (8); Methanamine, 1,1-dimethoxy-N,N-dimethyl- (9); (4637-24-5)

(−)-SALSOLIDINE

(Isoquinoline, 1,2,3,4-tetrahydro-6,7-dimethoxy-1-methyl-, (S)-)

Submitted by Albert I. Meyers,[1] Michael Boes, and Daniel A. Dickman.
Checked by Melinda Gugelchuk and Leo A. Paquette.

1. Procedure

A. *6,7-Dimethoxy-1,2,3,4-tetrahydro-2-[(1-tert-butoxy-3-methyl)-2-butyl-iminomethyl]isoquinoline* (**2**). In a 250-mL, round-bottomed flask 10.0 g (51.7 mmol) of 6,7-dimethoxy-1,2,3,4-tetrahydroisoquinoline 1 (Note 1) is combined with 11.5 g (53.7 mmol) of (S)-N,N-dimethyl-N'-(1-tert-butoxy-3-methyl)-2-butylformamidine (Note 2), 50 mL of dry toluene and 50 mg of (+)-camphorsulfonic acid (Note 3). The mixture is heated to reflux for 24 hr and allowed to cool to room temperature. Approximately 30 mL of toluene is removed by rotary evaporation and the residual solution is heated at reflux for an additional 2 days. After the reaction mixture is cooled, it is diluted with 50 mL of dichloromethane and washed with 50 mL of 1 N sodium hydroxide and 100 mL of brine and the organic layer is dried over anhydrous potassium carbonate, filtered, and concentrated by rotary evaporation. The residue is distilled (Kugelrohr 0.1 mm, 170°C bath temp) to give 18.0 g (96%) of **2** as a pale yellow oil (Note 4).

B. *(S)-6,7-Dimethoxy-1-methyl-1,2,3,4-tetrahydroisoquinoline, (-)-salsolidine* (**3**). A 500-mL, three-necked flask, containing a magnetic stirring bar, is equipped with a three-way stopcock, low temperature thermometer, and a rubber septum. The flask is charged with 15.0 g (41.4 mmol) of formamidine **2**, filled with argon, and kept under a pressure of ca. 100 mm against the atmosphere (Note 5). Through the septum, via a syringe, is added 300 mL of dry tetrahydrofuran (Note 6) and the solution is cooled to -75°C in a dry ice-acetone bath. A tert-butyllithium solution (21 mL of a 2.4 M solution, Note 7), is added dropwise within 5 min through the septum. After the solution is stirred at -75°C for 45 min, the deep red solution is cooled to -100°C in a liquid nitrogen-methanol bath and, after 15 min at -100°C, 3 mL of freshly distilled iodomethane (Note 8) dissolved in 10 mL of dry tetrahydrofuran is added by syringe at such a rate that the temperature of the reaction mixture does not rise above -90°C. Stirring is continued for 3 hr and the solution is poured into a 1-L separatory funnel containing 50 mL of water. This is extracted twice with 100 mL of dichloromethane and the combined organic layers are washed with 100 mL of brine, dried over potassium carbonate, and filtered. Removal of the solvent on a rotary evaporator gives a cloudy yellow oil, which is dissolved in 100 mL of 60% ethanol. To this solution is added 4.5-5.0 mL of hydrazine (Note 9) followed by 3.0 mL of glacial acetic acid (pH 8-9). The mixture is stirred overnight at ambient temperature and diluted with 50 mL of water. It is extracted twice with 100 mL of dichloromethane and the combined organic extracts are washed with 50 mL of water, dried (potassium carbonate), filtered and concentrated at *ambient temperature under aspirator pressure*. The residue, which consists of valinol tert-butyl ether and salsolidine, is distilled bulb-to-bulb under aspirator pressure at 105°C (pot temperature). This removes the valinol tert-butyl ether (Note 10), leaving crude salsolidine as the pot residue. The residue is dissolved in 100 mL of ether and washed twice with 35-mL portions of ice water - 3 N hydrochloric acid (1:4). The ether layer is discarded and the acidic aqueous layer is neutralized with cold (0-5°C) aqueous 25% sodium hydroxide until it is alkaline to pH paper. The creamy mixture is immediately extracted twice with 50 mL of dichloromethane and the organic layers are drawn off, combined, and dried over anhydrous potassium carbonate. After the drying agent is removed by filtration, it is washed twice with 5 mL of dichloromethane. The filtrate and wash are concentrated by rotoevaporation, leaving a yellow oil. Distillation (Kugelrohr) at a pot temperature of 120-125°C (0.01 mm) gives 5.1-5.3 g (60-63%) of (S)-6,7-dimethoxy-1-methyl-1,2,3,4-tetrahydroisoquinoline (salsolidine) 3 as a pale yellow oil, which crystallizes on standing (Note 11), mp 47-49°C, (Notes 12 and 13).

2. Notes

1. The tetrahydroisoquinoline was purchased from Aldrich Chemical Company, Inc. and treated with aqueous 5% sodium hydroxide, extracted with dichloromethane, dried over sodium sulfate and distilled, bulb-to-bulb at 0.01 mm (58-60°C, bath temp) to give a colorless oil. The compound solidifies to give an amorphous solid: mp 83.0-84.5°C; ^{1}H NMR (CDCl$_3$, 300 MHz) δ: 2.30 (br s, 1 H), 2.69, (t, 2 H, J = 6), 3.12 (t, 2 H, J = 6), 3.80 (s, 3 H), 3.82 (s, 3 H), 3.95 (s, 2 H), 6.50 (s, 1 H), 6.58 (s, 1 H).

2. See the previous procedure for preparation of this chiral formamidine; *Org. Synth.* **1988**, *67*, 52.

3. (+)-10-Camphorsulfonic acid was purchased from Aldrich Chemical Company, Inc.

4. The physical properties are as follows: ^{1}H NMR (CDCl$_3$, 300 MHz) δ: 0.87 (d, 3 H, J = 6.8), 0.87 (d, 3 H, J = 6.7), 1.14 (s, 9 H), 1.83 (hept, 1 H, J = 7), 2.68-2.81 (m, 3 H), 3.20 (dd, 1 H, J_1 = 8.8, J_2 = 7.1), 3.43-3.56 (m, 3 H), 3.83 (s, 3 H), 3.85 (s, 3 H), 4.40 (AB, 1 H, J = 17), 4.43 (AB, 1 H, J = 17), 6.60 (s, 1 H), 6.63 (s, 1 H), 7.40 (s, 1 H); $[\alpha]_D^{25}$ -30.3° (CHCl$_3$, *c* 2.7).

5. The flask was filled with argon by evacuating and pressurizing several times through the three-way stopcock.

6. Tetrahydrofuran was distilled from sodium wire and benzophenone.

7. tert-Butyllithium solution in pentane was purchased from Alfa Products, Morton/Thiokol Inc.

8. Iodomethane was purchased from Aldrich Chemical Company, Inc.

9. Anhydrous hydrazine was purchased from Aldrich Chemical Company, Inc.

10. The spectral properties are as follows: ^{1}H NMR (CDCl$_3$) δ: 0.92 (d, 6 H, J = 6), 1.19 (s, 9 H), 1.61 (m, 1 H), 2.64 (m, 1 H), 3.12 (overlapping doublets, 1 H, J = 8, J = 9), 3.40 (dd, 1 H, J = 8, J = 9).

11. The checkers did not observe crystallization at this point when the reaction was run on a much smaller scale (8 mmol). Two runs, each starting with 3.05 g of 2, gave rise to 0.88 g (51%) and 0.83 g (46%) of purified (-)-salsolidine.

12. The physical properties are as follows: ^{1}H NMR (300 MHz, CDCl$_3$) δ: 1.41 (d, 3 H, J = 6.6), 1.78 (br s, 1 H), 2.65 (m, 1 H), 2.77 (m, 1 H), 3.01 (m, 1 H), 3.25 (m, 1 H), 3.822 (s, 3 H), 3.828 (s, 3 H), 3.89 (m, 1 H), 6.61 (s, 1 H), 6.55 (s, 1 H); $[\alpha]_D^{22}$ -53.9 to -54.0° (C$_2$H$_5$OH, c 3.8) (lit.^{2} -59.5° ± 0.5°); (95-95% ee by Pirkle HPLC analysis, Note 13).

13. Enantiomeric purity determination was performed as follows: The (-)-salsolidine 3 (100 mg) was dissolved in 1 mL of dichloromethane and 0.05 mL of triethylamine. To this solution was added 150 mg of 1-naphthoyl chloride (Aldrich Chemical Company, Inc.) and the reaction mixture was stirred for 0.5 hr at room temperature, then poured into 5 mL of aqueous 20% sodium hydroxide solution. Extraction with 10 mL of dichloromethane was followed by separation of the organic layer which was dried over potassium carbonate. The solvent was removed under reduced pressure and the crude naphthamide was purified by column chromatography on silica gel (ethyl acetate-hexane-dichloromethane, 1:4:5). Enantiomeric analysis was performed using a Waters Associates Model 440 high pressure liquid chromatograph equipped with a Pirkle Covalent Phenylglycine Column (Baker Bond Chiral HPLC Column, DNBPG, J. T. Baker, Phillipsburg, NJ). The detector used was UV at 254 nm and the solvent for elution was 10% 2-propanol-hexane at 5 mL/min with a back pressure of 3500 psi. The peaks were baseline separated and electronically integrated.

3. Discussion

This method of asymmetric alkylation has been performed in a number of other systems with equally good enantioselectivity. Tetrahydroisoquinolines have been alkylated[3] (Eq 1) with various alkyl halides to give 1-substituted tetrahydroisoquinolines in 50-70% overall yields and with excellent ee's. Several naturally occurring isoquinoline alkaloids have also been prepared (compounds A-C) in 95-98.5% ee.[4] A number of chiral auxiliaries other than the valine-based tert-butyl ether also have been examined and gave 80-99% ee's after alkylation.[5] However, the authors consider the chiral auxiliary used in the present procedure to be superior to the others.

TABLE I

R,R	R'X	% ee
H	MeI	99
H	n-BuI	96
H	Allyl Br	96
H	PhCH$_2$Cl	98
MeO, MeO	3,4-(MeO)$_2$PhCH$_2$Br	99
MeO, MeO	3,4-(MeO)$_2$PhCH$_2$CH$_2$I	95

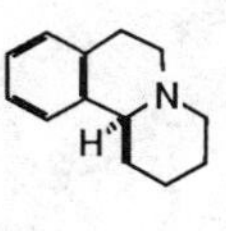

In addition to isoquinolines, β-carbolines have been used to afford indole alkaloids, both natural and unnatural (Scheme 1) in high enantiomeric excess.[6] The indole nitrogen is protected as the methoxymethyl ether and later removed to provide the unsubstituted indole. In the absence of indole-nitrogen protection, the potassium-salt is satisfactory but results in low asymmetric induction. However, if racemic products are desired, N-tert-butyl-formamidines can be used,[7] and smooth alkylation of the α-protons is achieved, thus obviating the need for protection of the indole nitrogen.

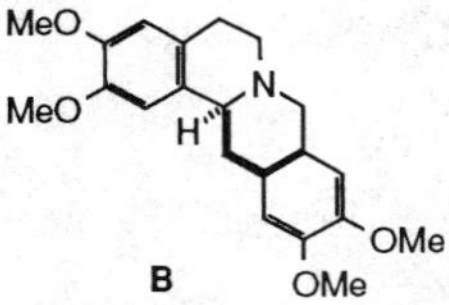

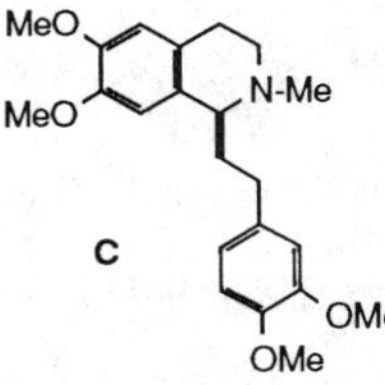

Scheme 1

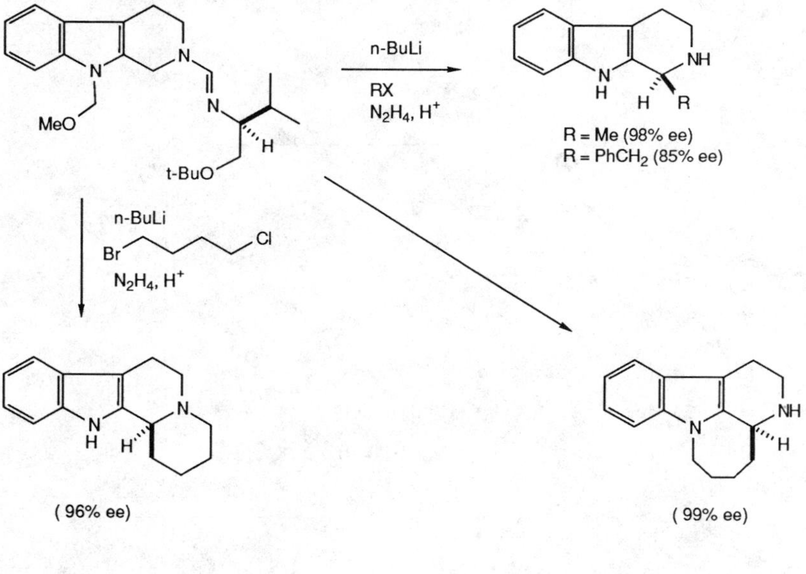

Asymmetric alkylations also are feasible, leading to the chiral dihydropyrrole (D) and the tetrahydropiperidine systems (E).[8] When the

saturated analogs were employed (pyrrolidine and piperidine), no metallation could be effected in the presence of the chiral auxiliary, although metallation-alkylation proceeded normally when the N-tert-butylformamidines were employed.

1. Department of Chemistry, Colorado State University, Fort Collins, CO 80523.
2. Battersby, A. R.; Edwards, T. P. *J. Chem. Soc.* **1960**, 1214.
3. Meyers, A. I.; Boes, M.; Dickman, D. A. *Angew. Chem., Intern. Ed. Engl.* **1984**, *23*, 458; Dickman, D. A.; Meyers, A. I. *Tetrahedron Lett.* **1986**, *27*, 1465.
4. For a recent report on the scope of this method in isoquinoline alkaloid syntheses, see Meyers, A. I.; Dickman, D. A.; Boes, M. *Tetrahedron* **1987**, *43*, 5095.

5. Meyers, A. I.; Fuentes, L. M. *J. Am. Chem. Soc.* **1983**, *105*, 117; Meyers, A. I.; Fuentes, L. M.; Kubota, Y. *Tetrahedron* **1984**, *40*, 1361.
6. Meyers, A. I.; Loewe, M. F. *Tetrahedron Lett.* **1985**, *26*, 3291; Meyers, A. I.; Sohda, T.; Loewe, M. F. *J. Org. Chem.* **1986**, *51*, 3108.
7. Meyers, A. I.; Loewe, M. F. *Tetrahedron Lett.* **1984**, *25*, 2641; Meyers, A. I.; Edwards, P. D.; Rieker, W. F.; Bailey, T. R. *J. Am. Chem. Soc.* **1984**, *106*, 3270; Meyers, A. I.; Hellring, S.; Hoeve, W. T. *Tetrahedron Lett.* **1981**, *22*, 5115.
8. Meyers, A. I.; Dickman, D. A.; Bailey, T. R. *J. Am. Chem. Soc.* **1985**, *107*, 7974.

Appendix

Chemical Abstracts Nomenclature (Collective Index Number); (Registry Number)

(-)-Salsolidine: Salsolidine (8); Isoquinoline, 1,2,3,4-tetrahydro-6,7-dimethoxy-1-methyl-, (S)- (9); (493-48-1)

6,7-Dimethoxy-1,2,3,4-tetrahydro-2-[(1-tert-butoxy-3-methyl)-2-butylimino-methyl]isoquinoline: Isoquinoline, 2-[[[1-[(1,1-dimethylethoxy)methyl]-2-methylpropyl]imino]methyl]-1,2,3,4-tetrahydro-6,7-dimethoxy-, (S)- (11); (90482-03-4)

6,7-Dimethoxy-1,2,3,4-tetrahydroisoquinoline hydrochloride: Isoquinoline, 1,2,3,4-tetrahydro-6,7-dimethoxy-, hydrochloride (8,9); (2328-12-3)

(S)-N,N-Dimethyl-N'-(1-tert-butoxy-3-methyl-2-butyl)formamidine: Methanimidamide, N'-[1-[(1,1-dimethylethoxy)methyl]-2-methylpropyl]-N,N-dimethyl-, (S)- (11); (90482-06-7)

N-tert-BUTOXYCARBONYL-L-LEUCINAL

(Carbamic acid, (1-formyl-3-methylbutyl)-,
1,1-dimethylethyl ester, (S)-)

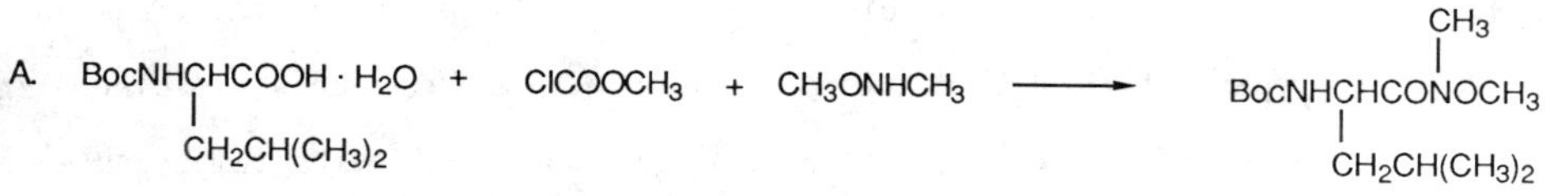

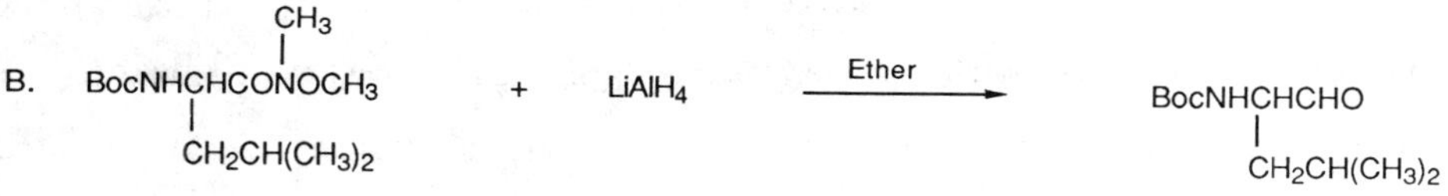

A. BocNHCHCOOH · H₂O + ClCOOCH₃ + CH₃ONHCH₃ ⟶ BocNHCHCON(CH₃)OCH₃
      | |
    CH₂CH(CH₃)₂   CH₂CH(CH₃)₂

B. BocNHCH(CH₂CH(CH₃)₂)CON(CH₃)OCH₃ + LiAlH₄ —Ether→ BocNHCH(CH₂CH(CH₃)₂)CHO

Submitted by O. P. Goel, U. Krolls, M. Stier, and S. Kesten.[1]
Checked by Susumu Ohira and James D. White.

1. Procedure

A. *Boc-L-Leucine N-methyl-O-methylcarboxamide.* A 1-L, three-necked, round-bottomed flask is equipped with a mechanical stirrer, an electronic digital thermometer, and a graduated addition funnel. The flask is charged with 39.1 g (0.4 mol) of N,O-dimethylhydroxylamine hydrochloride (Note 1) and 236 mL of methylene chloride (Note 2). The suspension is stirred and cooled to 2°C with an ice-water bath. N-Methylpiperidine (Note 3), 48.8 mL (0.41 mol), is placed in the addition funnel and added dropwise while the temperature is maintained at 2° ± 2°C. A clear, colorless solution results which is kept cold and used in the following reaction.

A 5-L, three-necked, round-bottomed flask is equipped with a mechanical stirrer, thermometer, and an addition funnel with drying tube. The flask is charged with 100 g (0.4 mol) of Boc-L-leucine hydrate (Note 4), 458 mL of tetrahydrofuran (Note 2), and 1.8 L of methylene chloride. A clear solution results on stirring, which is cooled to -20° ± 2°C by immersing the flask in a dry ice-2-propanol bath. N-Methylpiperidine, 48.8 mL (0.41 mol), is placed in the addition funnel and added rapidly to the mixture, while the temperature is allowed to rise to -12° ± 2°C. Methyl chloroformate (Note 5), 31 mL (0.4 mol) is then placed in the addition funnel and added rapidly to the mixture with good stirring, while the temperature is kept at -12° ± 2°C. Two minutes later the solution of N,O-dimethylhydroxylamine, prepared as described earlier, is added. The cooling bath is removed and the clear solution allowed to warm to room temperature over 4 hr (Note 6). The solution is again cooled to 5°C and extracted with two 500-mL portions of aqueous 0.2 N hydrochloric acid and two 500-mL portions of aqueous 0.5 N sodium hydroxide (Note 7). The solution is washed with 500 mL of saturated aqueous sodium chloride solution, dried over magnesium sulfate, and concentrated on a rotary evaporator at a bath temperature of 30-35°C. The residue is further evacuated on an oil pump to constant weight. The residual colorless syrup weighs 100-102 g (91-93%), $[\alpha]_D^{23}$ -24 to -25° (1.5% in methanol) (Note 8).

B. *N-tert-Butoxycarbonyl-L-leucinal: Boc-L-leucinal.* A 5-L, four-necked, round-bottomed flask is equipped with an efficient mechanical stirrer, a thermometer, a pressure-equalizing addition funnel, and an air-cooled condenser fitted with an argon blanket adapter. The flask is charged under an argon blanket with 17.7 g (95% pure, 0.44 mol) of lithium aluminum hydride (Note 9), and 1.5 L of anhydrous ethyl ether (Note 10). The grey suspension is stirred at room temperature for 1 hr or until most of the solid is finely dispersed. The flask is immersed in a dry ice-2-propanol bath and the suspension cooled to -45°C. A solution of the Boc-L-leucine N-methyl-O-methylcarboxamide, obtained in Part A, in 300 mL of anhydrous ethyl ether is placed in the addition funnel and added to the lithium aluminum hydride suspension in a steady stream (Note 11) while the reaction temperature is maintained -35° ± 3°C. The cooling bath is removed and the mixture is stirred and allowed to warm to +5°C. The mixture is once again cooled to -35°C and a solution of 96.4 g (0.71 mol) of potassium bisulfate (Note 12) in 265 mL of deionized water is placed in the addition funnel. This is added cautiously at first and then rapidly, while the temperature is allowed to rise to -2° ± 3°C. The cooling bath is removed and the mixture stirred for 1 hr. The reaction mixture is filtered through a 2" pad of Celite (Note 13). The filter cake is washed with two 500-mL portions of ethyl ether. The combined ether layers are washed in sequence with three 350-mL portions of cold (5°C) 1 N hydrochloric acid, two 350-mL portions of saturated aqueous sodium bicarbonate solution, and 350 mL of saturated sodium chloride solution. The organic solution is dried over magnesium sulfate and evaporated on a rotary evaporator (bath at 30°C). The residual, slightly cloudy syrup weighs 69-70 g (87-88%), $[\alpha]_D^{23}$ -49 to -51° (1.65% in methanol) (Note 14). The product is stored in a freezer (-17°C) prior to use. It solidifies readily at 5°C (Note 15).

2. Notes

1. N,O-Dimethylhydroxylamine hydrochloride was obtained from the Aldrich Chemical Company, Inc., and used as received.

2. Methylene chloride and tetrahydrofuran were obtained from the Fisher Scientific Company.

3. N-Methylpiperidine was obtained from the Aldrich Chemical Company, Inc., and used as received.

4. Bachem Inc. was the source of Boc-L-leucine hydrate. It was not necessary to prepare anhydrous Boc-L-leucine. The yield and quality of the product were unaffected by the presence of water during the reaction.

5. Methyl chloroformate was obtained from the Aldrich Chemical Company, Inc.

6. The reaction mixture may be stirred overnight for convenience.

7. The organic solution should be kept at 5° to 15°C during extractions.

8. The crude product is 96-99% pure by HPLC and is satisfactory for use in the next reaction. HPLC was carried out on a Varian 5500 instrument using a 250 cm x 4.6 mm I.D. Alltech C-18 column with 60:40 methanol:0.5 M $NH_4H_2PO_4$ (pH 3) as the mobile phase, UV detector at 210 nm. Thin layer chromatography on silica gel plates (EM) and development with hexane:EtOAc(3:1) indicates that the major product spot is at R_f = 0.32. Starting Boc-L-leucine, if present, appears at R_f = 0.16; detector: ninhydrin and gradual warming on a hot plate. The physical properties are as follows: IR (liquid film) cm^{-1}: 2960(s), 1714(s), 1665(s); ^1H NMR (200 MHz, $CDCl_3$) δ: 0.92 (2d appear as t, 6 H, J = 6.8, $CH(CH_3)_2$), 1.40 (s, 9 H, $C(CH_3)_3$), 1.38-1.44 (m, 2 H, C_3-H), 1.59-1.76 (m, 1 H, C_4-H), 3.17 and 3.14 (s and a rotamer singlet, 3 H, N-CH_3), 3.76 and 3.67 (s, and a rotamer singlet, 3 H, O-CH_3), 4.7 (m, 1 H, C_2-H), 5.06 (m, 1 H, N-H).

9. Lithium aluminum hydride was obtained from Alfa Products, Morton/Thiokol Inc.

10. Ethyl ether was obtained from the Fisher Scientific Co.

11. The lithium aluminum hydride suspension should be cooled to -45°C prior to addition of the amide. Higher initial temperatures (-30°C and above) lead to an impurity as shown by TLC.

12. Potassium bisulfate was obtained from the Matheson, Coleman and Bell Co. A saturated aqueous solution is obtained after stirring overnight. Aqueous potassium bisulfate will react vigorously if tetrahydrofuran is the reaction medium in place of ethyl ether.

13. A gel-like precipitate is formed from the inorganic by-products. A thick Celite pad helps to prevent clogging of the filter funnel.

14. Consistently higher optical rotations than reported were obtained.[2] NMR and capillary gas chromatographic analyses indicated chemical purity of 98 to 99%. Varian 6000, RSL-310, 15 meter, fused silica column, 0.25 mm ID, film thickness 0.25 μm, at 60°C for 4 min and then 60-220°C at 10°C/min, H_2 as carrier gas at 10 psi. TLC under the conditions described in Note 8 shows the major spot at R_f = 0.53. The spectral properties are as follows: IR (liquid film) cm^{-1}: 2961(s), 1736(s), 1698(b); ^{1}H NMR (200 MHz, $CDCl_3$) δ: 0.96 (d, 6 H, J = 6.4, CH $(CH_3)_2$), 1.45 (s, 9 H, $C(CH_3)_3$), 1.48-1.81 (m, 3 H, C_3-H, C_4-H); 4.24 (m, 1 H, C_2-H); 4.92 (broad singlet, 1 H, N-H); 9.59 (s, 1 H, C_1-H).

15. Boc-L-Leucinal racemizes if stored at room temperature. Although it solidified in the cold it became liquid at room temperature. It is very soluble in pentane at room temperature, but crystallizes from it at -30°C. It is reported[2] to melt at 63-66°C.

3. Discussion

Boc-L-Leucinal is a useful chiral synthon in the preparation of the natural amino acid statine[3] [S-(R*,R*)]-4-amino-3-hydroxy-6-methylheptanoic acid (3S,4S). The procedure reported here is based on the method of Fehrentz and Castro[2] for the preparation of optically active Boc amino aldehydes from α-amino acids. It is satisfactory on a kilogram scale. Boc-L-Leucinal has also been prepared by the reduction of Boc-L-leucine methyl ester with diisobutylaluminum hydride[4] or by oxidation of Boc-L-leucinol.[5] The reaction conditions described here differ from those in the literature.[2] The N-methoxy-N-methylamide is prepared simply and in high yield by the mixed anhydride method[6] rather than with the very expensive reagent benzotriazol-1-yloxytris(dimethylamino)phosphonium hexafluorophosphate. In addition, the amide can be added to a cold lithium aluminum hydride suspension rather than inversely as recommended.[2] This is an important consideration for scale-up. Reduction of this amide with bis(2-methoxyethoxy)aluminum hydride solution (Red-Al) gave a substantially impure aldehyde.

1. Warner-Lambert Company, Parke-Davis Pharmaceutical Research, 2800 Plymouth Road, Ann Arbor, MI 48105.
2. Fehrentz, J.-A.; Castro, B. *Synthesis* **1983**, 676.
3. Woo, P. W. K. *Tetrahedron Lett.* **1985**, *26*, 2973 and references cited therein.
4. Rich, D. H.; Sun, E. T.; Boparai, A. S. *J. Org. Chem.* **1978**, *43*, 3624.
5. Rittle, K. E.; Homnick, C. F.; Ponticello, G. S.; Evans, B. E. *J. Org. Chem.* **1982**, *47*, 3016.

6. Chen, F. M. F.; Steinauer, R.; Benoiton, N. L. *J. Org. Chem.* **1983**, *48*, 2939.

Appendix
Chemical Abstracts Nomenclature (Collective Index Number);
(Registry Number)

N-tert-Butoxycarbonyl-L-leucinal: Carbamic acid, (1-formyl-3-methylbutyl)-, 1,1-dimethylethyl ester, (S)- (9); (58521-45-2)

N-tert-Butoxycarbonyl-L-leucine N-methyl-O-methylcarboxamide: Carbamic acid, [1-[(methoxymethylamino)carbonyl]-3-methylbutyl]-, 1,1-dimethylethyl ester (S); (11); (87694-50-6)

N,O-Dimethylhydroxylamine hydrochloride: Methylamine, N-methoxy-, hydrochloride; (8); Methanamine, N-methoxy-, hydrochloride (9); (6638-79-5)

N-Methylpiperidine: Piperidine, 1-methyl- (8,9); (626-67-5)

N-tert-Butoxycarbonyl-L-leucine hydrate: Leucine, N-carboxy-, N-tert-butyl ester, L- (8); L-Leucine, N-[(1,1-dimethylethoxy)carbonyl]- (9); (13139-15-6)

CONDENSATION OF (-)-DIMENTHYL SUCCINATE DIANION WITH 1,ω-DIHALIDES: (+)-(1S,2S)-CYCLOPROPANE-1,2-DICARBOXYLIC ACID

(1,2-Cyclopropanedicarboxylic acid, (1S,1S)-(+)-)

A. [succinic anhydride] + [(-)-menthol] →(TsOH, toluene)→ dimenthyl succinate (CO$_2$Menthyl, CO$_2$Menthyl)

B. dimenthyl succinate →(1. LTMP, THF; 2. BrCH$_2$Cl)→ cyclopropane diester (H, CO$_2$Menthyl, H, CO$_2$Menthyl)

C. cyclopropane diester →(1. KOH, MeOH, H$_2$O; 2. HCl)→ cyclopropane diacid (H, CO$_2$H, H, CO$_2$H)

Submitted by Kyoji Furuta, Kiyoshi Iwanaga, and Hisashi Yamamoto.[1]
Checked by Ichiro Mori and Clayton H. Heathcock.

1. Procedure

A. *(-)-Dimenthyl succinate.* A 300-mL, one-necked, round-bottomed flask is equipped with a magnetic stirrer, Dean-Stark trap, and a reflux condenser. The flask is charged with 20 g (0.2 mol) of succinic anhydride,

62.5 g (0.4 mol) of ℓ-menthol, 250 mg (1.3 mmol) of p-toluenesulfonic acid monohydrate, and 150 mL of toluene (Note 1). The mixture is heated under reflux in an oil bath (about 140°C) for 24 hr. During this period the theoretical amount of water (3.6 mL) is collected. The mixture is allowed to cool to ambient temperature, diluted with 200 mL of hexane, and poured into a mixture of 250 mL of aqueous saturated sodium bicarbonate, 100 mL of methanol, and 200 mL of water. The organic phase is separated and the aqueous phase is extracted twice with 100 mL of hexane. The organic phases are combined, washed once with 200 mL of saturated brine, and dried over sodium sulfate. The solvent is removed with a rotary evaporator, and the resulting crude product is dissolved in 100 mL methanol. After the solution stands in a refrigerator overnight, colorless crystals appear in the mixture and are collected by filtration with suction (Note 2). This material (ca. 70 g in two crops) is purified by recrystallization from methanol to afford 66 g (84%) of pure (-)-dimenthyl succinate, mp 63-64°C (Note 3).

B. *(-)-Dimenthyl (1S,2S)-cyclopropane-1,2-dicarboxylate.* A dry, 500-mL, three-necked, round-bottomed flask containing a magnetic stirring bar is equipped with a 100-mL, pressure-equalizing dropping funnel, rubber septum, and a three-way stopcock with a nitrogen inlet. The air in the system is replaced with dry nitrogen. The flask is charged with 180 mL of dry tetrahydrofuran and cooled with an ice bath; 74.1 mL of a 1.7 M hexane solution of butyllithium (126 mmol) (Note 4) is added. This solution is stirred while 21.3 mL (126 mmol) of 2,2,6,6-tetramethylpiperidine (Note 5) is added dropwise with a syringe through the septum over a 10-min period. The resulting solution of lithium 2,2,6,6-tetramethylpiperidide (LTMP) is cooled to -78°C with a dry ice-methanol bath (Note 6) and stirred. A solution of 23.7 g (60 mmol) of (-)-dimenthyl succinate in 50 mL of dry tetrahydrofuran is then added dropwise through the addition funnel over a period of 1 hr. The wall of the funnel is rinsed with 10 mL of dry tetrahydrofuran and the rinse is added to the solution. The resulting yellow solution of succinate dianion is stirred for 1 hr. To the solution is added dropwise 3.9 mL (60 mmol) of bromochloromethane (Note 7) with a syringe through the septum over a 10-min period. After the reaction mixture is stirred for 2 hr (Note 8), 2.2 mL (24 mmol) of isobutyraldehyde (Note 9) is added dropwise to quench any unreacted anions (Note 10). After being stirred for an additional 30 min, the reaction mixture is poured into 250 mL of ice-cooled 1 N hydrochloric acid and the product is extracted three times with 150 mL of ether. The combined organic phases are washed with 250 mL of brine, dried over sodium sulfate, filtered, and concentrated with a rotary evaporator. The residue is chromatographed on 700 g of silica gel (Note 11) packed in a 9.5-cm diameter column using a mixture of ether and hexane (1:18) as eluant. The appropriate fractions are collected and concentrated to give 11.5 g (47%) of (-)-dimenthyl (1S,2S)-cyclopropane-1,2-dicarboxylate as colorless crystals. Analysis by GC indicates a diastereomeric ratio of 96:4 (Note 12). Recrystallization of this material from 25 mL of methanol affords 9-10 g (38-40%) (Note 13) of optically pure product, mp 95-96°C (Note 14).

C.   *(+)-(1S,2S)-Cyclopropane-1,2-dicarboxylic acid.*   (-)-Dimenthyl (1S,2S)-cyclopropane-1,2-dicarboxylate (4.06 g, 10 mmol, $[\alpha]_D^{25}$ +17.8°) is dissolved in 20 mL of a 10% potassium hydroxide solution in 9:1 methanol/water in a 50-mL, one-necked, round-bottomed flask equipped with a magnetic stirring bar. The solution is heated at 60°C with an oil bath. Progress of the reaction is monitored by TLC on silica gel, using 1:1 hexane/ethyl acetate as eluant (Note 15). After about 4 hr the resulting two-phase mixture is diluted with 20 mL of water and extracted with three 40-mL portions of ether (Note

16). The aqueous layer is acidified by the addition of 20 mL of ice-cold 6 N hydrochloric acid, saturated with ca. 5 g of sodium chloride, and extracted with five 40-mL portions of ether. The combined organic layers are dried over sodium sulfate, diluted with 5 mL of hexane, and concentrated with a rotary evaporator. Filtration provides 1.17 g (90%) of (+)-(1S,2S)-cyclopropane-1,2-dicarboxylic acid, mp 172-173°C, $[\alpha]_D^{25}$ +228° (ethanol, c 1.01) [lit.^{2} mp 169.5-170°C, $[\alpha]_D^{25}$ +227.9° (ethanol, c 2.342)] (Note 17).

2. Notes

1. Succinic anhydride and p-toluenesulfonic acid monohydrate were purchased from Wako Pure Chemical Industries, Ltd. (Japan). Guaranteed-grade ℓ-(-)-menthol was purchased from Tokyo Kasei Kogyo Company, Ltd. (Japan). Reagent-grade toluene was dried and stored over sodium metal. The checkers obtained p-toluenesulfonic acid from Eastman Kodak, and succinic anhydride and ℓ-(-)-menthol from the Aldrich Chemical Company, Inc.

2. In the first trial the checkers experienced difficulty in crystallization at this point, even after keeping the solution in a 5°C-refrigerator for four days. Crystallization was induced by cooling the crude product (ca. 80 g of oil) to -78°C in a methanol-dry ice bath. Ether (5 mL) was added to the resulting glass, the cooling bath was removed, and the surface of the solid material was scratched continuously with a spatula. At about 0°C the glass began to melt and small white spots appeared. Continued stirring of the viscous material as it warmed resulted in crystallization of the entire mass. The crystalline mass was dried under vacuum and a small portion kept as seed crystals. The remainder was dissolved in 100 mL of warm methanol. After the solution was cooled to room temperature, a small quantity

of the seed material was added and the solution was placed in a 5°C-refrigerator overnight. Approximately 68 g of crystalline (-)-dimenthyl succinate was obtained. The filtrate was condensed with a rotary evaporator and cooled again to give another 2.6 g. In subsequent trials, the foregoing procedure was not necessary as the crude diester crystallized spontaneously, giving a similar yield in two crops.

3. The submitters report mp 65-66°C. The physical properties are as follows: ^{1}H NMR (CDCl$_3$, 250 MHz) δ: 0.70-2.02 (complex, 18 H), 0.75 (d, 6 H, J = 6.9), 0.89 (d, 12 H, J = 6.4), 2.60 (s, 4 H), 4.70 (dt, 2 H, J = 4.4, 10.8); $[\alpha]_D^{25}$ -88.7° (CHCl$_3$, c 1.02).

4. Tetrahydrofuran was freshly distilled from sodium-benzophenone. Butyllithium was obtained from Wako Pure Chemical Industries, Ltd. or Foote Mineral Company. It was titrated with anhydrous 2-butanol using 1,10-phenanthroline as an indicator.

5. 2,2,6,6-Tetramethylpiperidine, purchased from Tokyo Kasei Kogyo Company, Ltd. or Aldrich Chemical Company, Inc., was used. The use of this sterically-hindered lithium amide is crucial for high diastereoselectivity. If lithium diisopropylamide is used, the diastereoselectivity of the reaction is reduced significantly.[3]

6. The flask was cooled with a dry ice-methanol bath for 30 min before subsequent addition.

7. Bromochloromethane was purchased from Tokyo Kasei Kogyo Company, Ltd. or from Aldrich Chemical Company, Inc. and was used without purification.

8. TLC (ether-hexane) showed residual starting material.

9. Isobutyraldehyde was obtained from Wako Pure Chemical Industries, Ltd. or from Aldrich Chemical Company, Inc. and was used without purification.

10. This procedure was not essential but facilitated the subsequent chromatographic separation of desirable product from residue.

11. Silica gel (70-200 mesh) purchased from Fuji Davison Chemical (BW-820 MH) was used. The checkers used silica gel (230-400 mesh) purchased from Merck (Kieselgel 60).

12. GC analysis was performed with a capillary column (PEG, 0.25 mm x 25 m) purchased from Gaskuro Kogyo Company, Ltd. (Japan). The checkers used a 0.2 mm x 25 m Carbowax 20 M capillary column. The diastereomeric excess ranged from about 80 to 92%. In all cases, however, essentially pure material could be obtained by a subsequent single recrystallization. The diastereomeric ratio can be further improved by reducing the amount of added bromochloromethane. This may be due to the stereochemical purity of the enolates. On treatment with lithium 2,2,6,6-tetramethylpiperidide, the succinate affords mainly the E,E-enolate and only a slight amount of Z,Z-enolate; the latter may induce the opposite stereochemistry in the product. Fortunately, the E,E-enolate is more reactive, and therefore, the undesirable reaction with Z,Z-enolate can be suppressed kinetically by reducing the amount of halide. Thus the use of 0.5 equivalent of bromochloromethane compared to starting ester results in over 99% diastereoselectivity with comparable chemical yield, based on the halide. The procedure described in the text is, however, recommended from a practical point of view.

13. This yield ranges from 38 to 57% in several experiments. Recrystallization was accomplished by dissolving the crude diester in warm methanol and then allowing this solution to cool to room temperature. The checkers found that, if crystallization was induced by cooling the methanol solution in

a refrigerator, the crystalline diester was accompanied by an oily by-product. A second recrystallization was then necessary to purify the material.

14. The submitters report mp 99-100°C. The physical properties are as follows: ^{1}H NMR (CDCl$_3$, 250 MHz) δ: 0.70-2.02 (complex, 20 H), 0.76 (d, 6 H, J = 7.0), 0.90 (d, 12 H, J = 6.8), 2.12. (dd, 2 H, J = 7.4, 7.2), 4.68 (dt, 2 H, J = 5.4, 10.9); $[\alpha]_D^{25.5}$ +17.8° (CHCl$_3$, c 1.0).

15. During the course of the saponification, a spot with R_f intermediate between that of the reactant and product (persumably the monoester) was observed.

16. The combined organic layer may be washed with 50 mL of brine, dried over sodium sulfate, and evaporated to recover 3.45 g of crude menthol.

17. The checkers note that in the paper by Inouye, et al.[2] the diacid was prepared by ozonolysis of (+)-trans-2-phenylcyclopropanecarboxylic acid having 96.3% optical purity. The product diacid was purified by sublimation and recrystallization from water to obtain material giving the cited physical properties. Although the original publication[2] claims an optical purity of 96.3% for this diacid, it is probably optically pure because of the recrystallization step.

3. Discussion

The procedure described here provides a simple and general method for the construction of optically active trans-cycloalkane-1,2-dicarboxylic acids.[3] The reaction has been applied successfully to a series of dihalides and ditosylates (Table).

A few methods are described in the literature for the preparation of optically-active dialkyl trans-cyclopropane-1,2-dicarboxylates,[2] including a Michael addition-condensation sequence of menthyl chloroacetate and menthyl acrylate,[4] and cobalt(0) or nickel(0) complex-catalyzed cyclopropanation of dimenthyl fumarate with dibromomethane.[5] The present method is characterized by good chemical and high optical yields, simple operation, preparation of both enantiomers with equal ease, and the ready availability of the starting materials.

1. Department of Applied Chemistry, Faculty of Engineering, Nagoya University, Chikusa, Nagoya 464, Japan.
2. Inouye, Y.; Sugita, T.; Walborsky, H. M. *Tetrahedron* **1964**, *20*, 1695.
3. Misumi, A.; Iwanaga, K.; Furuta, K.; Yamamoto, H. *J. Am. Chem. Soc.* **1985**, *107*, 3343; Djerassi, C.; Klyne, W.; Norin, T.; Ohloff, G.; Klein, E. *Tetrahedron* **1965**, *21*, 163; Inouye, Y.; Sawada, S; Ohno, M.; Walborsky, H. M. *Tetrahedron* **1967**, *23*, 3237.
4. Inouye, Y.; Inamasu, S.; Ohno, M.; Sugita, T.; Walborsky, H. M. *J. Am. Chem. Soc.* **1961**, *83*, 2962; Inouye, Y.; Inamasu, S.; Horiike, M.; Ohno, M.; Walborsky, H. M. *Tetrahedron* **1968**, *24*, 2907; Inamasu, S.; Horiike, M.; Inouye, Y. *Bull. Chem. Soc. Jpn.* **1969**, *42*, 1393.
5. Matsuda, H.; Kanai, H. *Chem. Lett.* **1981**, 395.

TABLE

ASYMMETRIC SYNTHESIS OF CYCLOALKANE DICARBOXYLATES

$$(^-CHCOOR)_2 \;+\; (CH_2)_nX_2 \;\longrightarrow\; \underset{n+2}{\bigcirc}\overset{\text{\tiny\textbackslash\textbackslash\textbackslash COOR}}{\underset{COOR}{}}$$

Electrophile		Yield	% Diastereomeric	Configuration
X	n	(%)	Excess	
Br	2	72	_a,b	_c
Br	3	77	65	S,S
OTs	3	63	92	S,S
OTs	3	64	88	R,R[d]
Cl[e]	3	57	83[a]	_c
OTs	4	61	75	S,S

[a] Reaction first at -100°C, and then warming to -20°C.
[b] Not determined. $[\alpha]_D^{25}$ -51.7° (CHCl$_3$, c 1.0).
[c] Not determined.
[d] d-Menthyl ester was used.
[e] 3-Chloro-2-chloromethyl-1-propene.

Appendix

Chemical Abstracts Nomenclature (Collective Index Number);

(Registry Number)

(-)-Dimenthyl succinate: Butanedioic acid, bis[5-methyl-2-(1-methylethyl)-cyclohexyl]ester, [1R-[1α(1R*,2S*,5R*),2β,5α)]- (9); (34212-59-4)

(-)-Dimenthyl (1S,2S)-cyclopropane-1,2-dicarboxylate: 1,2-Cyclopropanedicarboxylic acid, bis[5-methyl-2-(1-methylethyl)cyclohexyl] ester [1S-[1 (1S*,2S*,5R*)], 2β,5α]]- (11); (96149-01-8)

-(-)-Menthol: Menthol, (-)- (8); Cyclohexanol, 5-methyl-2-(1-methylethyl)-, [1R-(1α,2β,5α)]- (9); (2216-51-5)

2,2,6,6-Tetramethylpiperidine: Piperidine, 2,2,6,6-tetramethyl- (8,9); (768-66-1)

(+)-(1S,2S)-Cyclopropane-1,2-dicarboxylic acid: 1,2-Cyclopropanedicarboxylic acid, (S,S)-(+)- (8); 1,2-Cyclopropanedicarboxylic acid, (1S-trans)- (9); (14590-54-6)

PALLADIUM-CATALYZED COUPLING OF ACID CHLORIDES WITH ORGANOTIN REAGENTS: ETHYL (E)-4-(4-NITROPHENYL)-4-OXO-2-BUTENOATE

(2-Butenoic acid, 4-(4-nitrophenyl)-4-oxo-, ethyl ester, (E)-)

A. Bu_3SnCl + $HC{\equiv}CLi \cdot NH_2CH_2CH_2NH_2$ $\xrightarrow{0° \text{ to } 25°C}$ $Bu_3Sn\text{-}C{\equiv}CH$

B. $Bu_3SnC{\equiv}CH$ + Bu_3SnH $\xrightarrow[90°C]{AIBN}$ $Bu_3Sn\text{-}CH{=}CH\text{-}SnBu_3$

C. $Bu_3Sn\text{-}CH{=}CH\text{-}SnBu_3$ $\xrightarrow{\substack{1)\ MeLi,\ -78°C \\ 2)\ ClCO_2Et,\ -78°C}}$ $Bu_3Sn\text{-}CH{=}CH\text{-}C(O)OEt$

D. $Bu_3Sn\text{-}CH{=}CH\text{-}C(O)OEt$ + 4-$O_2N\text{-}C_6H_4\text{-}C(O)Cl$ $\xrightarrow[Pd[0]]{CO,\ 50°C}$ 4-$O_2N\text{-}C_6H_4\text{-}C(O)\text{-}CH{=}CH\text{-}C(O)OEt$

Submitted by A. F. Renaldo, J. W. Labadie and J. K. Stille.[1]
Checked by Robert Aslanian, Cynthia A. Smith and Andrew S. Kende.

1. Procedure

Caution! Most tin compounds are toxic² and their preparation should be carried out in a well-ventilated hood.

A. *Tributylethynylstannane.* An oven-dried, 2-L, three-necked, round-bottomed flask equipped with a mechanical stirrer, 100-mL addition funnel, and a nitrogen inlet is charged with 24.0 g (0.26 mol) of lithium acetylide, ethylenediamine complex (Note 1). The system is evacuated, placed under nitrogen, and 800 mL of tetrahydrofuran (Note 2) is added to the system via a cannula. The flask is cooled in an ice water bath and 70.7 g (0.22 mol) of tributyltin chloride (Note 3) is added dropwise over 45 min. The ice bath is removed and the mixture is stirred for 18 hr at room temperature. The flask is placed in an ice water bath and excess lithium acetylide is hydrolyzed with 20 mL of water. The reaction mixture is concentrated under reduced pressure and washed with hexane (3 x 50 mL). The organic layers are combined and dried over anhydrous magnesium sulfate. Filtration and evaporation of the solvent at reduced pressure gives a colorless oil. Distillation yields 21.4-24.3 g (31-35%) of tributylethynylstannane, bp 90-94°C (0.5 mm) as a water-white liquid (Notes 4-6).

B. *(E)-1,2-Bis(tributylstannyl)ethylene.* In a 200-mL, one-necked, round-bottomed flask which contains a magnetic stirring bar and nitrogen inlet are placed 20.6 g (0.066 mol) of tributylethynylstannane, 23.1 g (0.079 mol) of tributyltin hydride (Note 7) and 0.25 g (0.0016 mol) of 2,2'-azobis(2-methylpropionitrile) (Note 8). The mixture is heated at 90°C with stirring for 6 hr. Distillation (170-186°C, 0.3 mm) yields 35.1-36.6 g (88-92%) of (E)-1,2-bis(tributylstannyl)ethylene as a clear colorless oil (Notes 9-10).

C. *Ethyl (E)-3-(tributylstannyl)propenoate*. A flame-dried, 1-L, three-necked, round-bottomed flask equipped with a magnetic stirring bar, 100-mL pressure-equalizing addition funnel (Note 11), and a nitrogen inlet is charged with 26.7 g (0.044 mol) of (E)-1,2-bis(tributylstannyl)ethylene. Tetrahydrofuran (100 mL) is added to the flask by cannula. The system is cooled in a dry ice-acetone bath and the addition funnel is charged, under nitrogen, with 44.8 mL of a 1.2 M solution of methyllithium (0.054 mol) in ethyl ether by means of a double-ended needle (Notes 12-13). After 10 min the lithium reagent is added dropwise to the flask over a period of 40 min. After the addition is complete, the yellow solution is stirred for an additional 2 hr at -78°C during which time a 1-L, one-necked, round-bottomed flask, containing a magnetic stirring bar, is flame-dried under nitrogen. The 1-L flask, capped with a rubber septum, is charged with a solution of 5.8 g (0.053 mol) of ethyl chloroformate (Note 14) in 150 mL of tetrahydrofuran and cooled to -78°C with a dry ice-acetone bath. Under gentle nitrogen pressure, the metallated reagent is transferred dropwise over a 2.0-hr period by means of a double-ended needle to the 1-L flask containing ethyl chloroformate while the temperature of both flasks is maintained at -78°C (Note 15). After the addition is complete, the reaction mixture is allowed to stir an additional 30 min at -78°C and then treated with 20 mL of methanol in one portion. After 10 min at -78°C the reaction mixture, while still cold, is transferred to a 1-L separatory funnel containing 200 mL of water and 100 mL of hexane. The organic layer is separated and the aqueous layer is washed with hexane (3 x 50 mL). The combined organic layers are dried over anhydrous sodium sulfate, filtered and concentrated to give a dark brown oil. The product is dissolved in hexane (30 mL) and purified by chromatography on a column of silica gel (600 g) (Note 16). Elution is carried out initially with hexane (Note 17) and then with hexane/ethyl acetate (95:5). Fractions containing the product are combined to give 10.2 g (59%) of ethyl (E)-3-(tributylstannyl)propenoate (Notes 18-20) as a yellow oil.

D. *Ethyl (E)-4-(4-nitrophenyl)-4-oxo-2-butenoate*. A flame-dried, 150-mL, one-necked, round-bottomed flask containing a magnetic stirring bar and equipped with a side-arm is charged with 3.20 g (17.2 mmol) of p-nitrobenzoyl chloride (Note 21), 0.08 g (0.10 mmol) of benzyl(chloro)bis(triphenylphosphine)palladium(II) (Note 22) and 30 mL of chloroform (Note 23). The bright yellow solution is evacuated and refilled with carbon monoxide (3 cycles) utilizing a gas bag (Notes 24-25). After an additional 10 min at room temperature a solution of 8.0 g (20.6 mmol) of ethyl (E)-3-(tributylstannyl)-propenoate in 5 mL of chloroform is added to the flask by syringe. The stirring reaction mixture is heated to 50°C for 12 hr while a pressure of 1 atm of carbon monoxide is maintained (Notes 26-27). The reaction is cooled to room temperature and treated with 18 mL of a 1.2 M solution of pyridinium poly(hydrogen fluoride) (Notes 28-30) along with 10 mL of pyridine. The reaction mixture is allowed to stir at room temperature overnight and then transferred to a 250-mL separatory funnel containing 75 mL of water. After addition of 30 mL of chloroform, the organic layer is washed successively with 10% hydrochloric acid (3 x 20 mL), saturated sodium bicarbonate (3 x 20 mL), water (25 mL), and brine (25 mL). The organic layer is dried over anhydrous sodium sulfate, filtered and concentrated to give a dark brown solid. The product is dissolved in chloroform and 15 g of silica gel (Note 16) is added to the solution. Concentration under reduced pressure gives a brown powder of silica coated with product, which is immediately placed on the top of a column of silica gel (50 g) (Note 16). Elution is carried out with ethyl acetate and the fractions are combined and concentrated under reduced pressure (Note

31). The crude product is again placed on a column of silica gel (250 g). Elution is carried out with hexane/ethyl acetate (90:10). Fractions containing the product (obtained by collecting the bright yellow band on the column) are combined to give 3.42 g (80%) of ethyl (E)-4-(4-nitrophenyl)-4-oxo-2-butenoate as yellow-green crystals, mp 69-71°C (Note 32).

2. Notes

1. Lithium acetylide, ethylenediamine complex is purchased from Aldrich Chemical Company, Inc. and used without purification.

2. Tetrahydrofuran is freshly distilled from sodium/benzophenone ketyl at atmospheric pressure under nitrogen.

3. Tributyltin chloride, purchased from Alfa Products, Morton/Thiokol, Inc., is distilled immediately before use (bp 128-130°C, 3 mm).

4. This procedure is a modification of that reported by Seitz.[3]

5. The remaining fraction of the mixture after distillation was bis-(tributylstannyl)acetylene which could be recycled for the preparation of tributylethynylstannane.

6. The spectral properties are as follows: ^{1}H NMR (270 MHz, CDCl$_3$) δ: 0.88 (t, 9 H, J = 7.3), 0.93-1.02 (m, 6 H), 1.25-1.38 (m, 6 H), 1.49-1.60 (m, 6 H), 2.17 (s, 1 H). The infrared spectrum (neat) shows absorption at 3260 and 2000 cm^{-1}.

7. Tributyltin hydride is prepared by the procedure of Hayashi[4] in 75% yield on a 0.3-mol scale. The checkers used material from Alfa Products, Morton/Thiokol, Inc., which was vacuum distilled before use (bp 75-78°C, 0.7 mm).

8. 2,2'-Azobis(2-methylpropionitrile), purchased from Alfa Products, Morton/Thiokol, Inc., is recrystallized from chloroform prior to use.

9. The submitters report bp 180-218°C (0.5 mm). The spectral properties are as follows: ^{1}H NMR (270 MHz, CDCl$_3$) δ: 0.75-1.02 (m, 30 H), 1.21-1.43 (m, 12 H), 1.48-1.63 (m, 12 H), 6.85 (s, 2 H). The infrared spectrum (neat) shows absorption at 1425 and 1020 cm^{-1}.

10. (E)-1,2-Bis(tributystannyl)ethylene has been prepared by an alternative procedure using lithium chloroacetylide.[5]

11. The funnel is capped with a rubber septum. For ease of operation, volume markings, which correspond to the amount of methyllithium to be added, are put on the addition funnel.

12. *Caution! Methyllithium is pyrophoric in air; excess quantities of the reagent should be discarded very carefully.*

13. Methyllithium is purchased from Aldrich Chemical Company, Inc. Although butyllithium could also be used in the metallation step, a cleaner product is obtained with methyllithium.

14. Ethyl chloroformate, purchased from Aldrich Chemical Company, Inc., is distilled at atmospheric pressure prior to use, discarding the first 25 mL.

15. The solution in the flask which contains the ethyl chloroformate is bright yellow and gradually becomes dark-red on the addition of the metallated reagent.

16. The checkers used Kieselgel 60, (230-400 mesh) purchased from E. Merck. The submitters used silica gel (32-63 mesh) purchased from Universal Scientific Inc.

17. Hexane removes the methyltributyltin.

18. The spectral properties are as follows: ^{1}H NMR (270 MHz, CDCl$_3$) δ: 0.78-0.99 (m, 12 H), 1.01-1.49 (m, 18 H), 4.11 (q, 2 H, J = 7.3), 6.22 (d, 1 H, J = 19.7), 7.65 (d, 1 H, J = 19.6). The infrared spectrum (neat) shows absorption at 1715, 1580 and 1200 cm^{-1}.

19. Attempts to purify the product by vacuum distillation, bp 110-138°C (0.05 mm), result in 7-8% isomerization to ethyl (Z)-3-(tributylstannyl)-propenoate [based on ^{1}H NMR (270 MHz) analysis].

20. The product should be stored under nitrogen at 0°C to prevent decomposition.

21. p-Nitrobenzoyl chloride, purchased from Aldrich Chemical Company, Inc., is recrystallized from hexane prior to use.

22. Benzyl(chloro)bis(triphenylphosphine)palladium(II) is prepared from tetrakis(triphenylphosphine)palladium(0)[6] (also available from Aldrich Chemical Company, Inc.) by the procedure of Fitton.[7]

23. Chloroform is freshly distilled at atmospheric pressure under nitrogen and filtered through a plug of neutral alumina.

24. The bright yellow color of the solution changes to light green after saturation with carbon monoxide. The presence of carbon monoxide prevents decarbonylation of the acylpalladium complex and thus the formation of ethyl p-nitrocinnamate.

25. The gas bag is purchased from Fisher Scientific.

26. The pressure of 1 atm is maintained by use of the gas bag.

27. The reaction changes color from light green to bright orange.

28. Pyridinium poly(hydrogen fluoride) is purchased from Aldrich Chemical Company, Inc.

29. The solution of pyridinium poly(hydrogen fluoride) in tetrahydrofuran and pyridine is prepared according to the procedure of Trost.[8] Pyridine is freshly distilled over calcium hydride at atmospheric pressure and stored over 4Å molecular sieves.

30. The orange reaction mixture changes to deep red and the reaction becomes slightly exothermic (50-60°C).

31. The initial filtration with silica gel is necessary to remove most of the tributyltin fluoride.

32. The spectral properties are as follows: ^{1}H NMR (270 MHz, CDCl$_3$) δ: 1.34 (t, 3 H, J = 7.2), 4.30 (q, 2 H, J = 7.2), 6.92 (d, 1 H, J = 15.6), 7.85 (d, 1 H, J = 15.6), 8.13 (d, 2 H, J = 8.9), 8.35 (d, 2 H, J = 8.8); ^{13}C NMR (68 MHz, CDCl$_3$) δ: 14.2, 61.7, 124.1, 129.9, 134.3, 135.5, 141.4, 151.0, 165.1, 188.4. The infrared spectrum (Nujol) shows the following absorption cm^{-1}: 1690, 1660, 1590, 1300, 990, 970 and 710.

3. Discussion

The procedure for synthesis of the title compound is representative of the palladium-catalyzed coupling of acid chlorides with organotin reagents.[9] The formation of ketones by Grignard reagents,[10] organocuprates,[11] and organoborates[12] has been reported. The principal advantage of the palladium-catalyzed organotin coupling method lies in the broad range of functionality that can be introduced in the product. The reaction can be carried out under mild, neutral conditions with functional groups on the acid chloride such as nitro, nitrile, haloaryl, methoxy, ester *and even aldehyde*.[9d] Solvents other than chloroform, such as tetrahydrofuran, hexamethylphosphorictriamide, and dichloromethane can be used in this reaction.

The unsymmetrical tetraorganotin reagent has been demonstrated to transfer selectively the vinyl group rapidly without butyl transfer occurring. By using a tributyl or trimethyl organotin reagent, e.g. Bu_3SnR or Me_3SnR, the order of transfer of the R groups is: $RC{\equiv}C-$ > $RCH{=}CH-$ > $Ar-$ > $RCH{=}CH-CH_2-$ > $ArCH_2-$ > CH_3OCH_2- > C_nH_{2n+1}.[13]

A number of functionalized organotin derivatives have been used in palladium-catalyzed coupling to produce aromatic heterocyclic ketones,[14] acetylenic ketones,[15] and vinyl ketones.[16] The organotin coupling method has been used effectively in the preparation of a key methyl ketone intermediate in the total synthesis of (±)-quadrone[17] and in the preparation of 5, a key precursor in the synthesis of the antibiotic pyrenophorin 6 (eq. 1).[13b]

The organotin reagents are very stable since they can withstand distillation as well as chromatography on silica gel. The procedure for

preparation of tributylethynylstannane (1) in Part A is based on one reported by Bottaro, Hanson and Seitz,[3] Bis(tributylstannyl)ethylene (2) has been prepared from lithium chloroacetylide[5] and tributylethynylstannane.[18] Although ethyl (E)-3-(tributylstannyl)propenoate (3) is produced from transmetallation[19] of 2 or hydrostannation[20] of ethyl propiolate, other known procedures to synthesize 3 include conjugate addition of tributylstannylcuprate[21] to ethyl propiolate, and tributylstannylcopper to β-substituted acrylates.[22]

In most cases the trimethyltin reagents are preferred since the by-product, trimethyltin chloride, can easily be removed by water extraction. In the case of the water-insoluble tributyltin chloride it is necessary to add an aqueous solution of potassium fluoride to an ethereal solution of the product thereby forming insoluble tributyltin fluoride, which can be separated by filtration.[13,23] However, a completely homogeneous and neutral fluoride source, pyridinium hydrofluoride,[8] is used in this procedure, making the filtration unnecessary and simplifying the subsequent chromatography step.

1. Department of Chemistry, Colorado State University, Ft. Collins, CO 80523.
2. (a) Luijten, J. G. A. In "Organotin Compounds"; Sawyer, A. K., Ed.; Marcel Dekker: New York, 1972; Vol. 3, pp. 949-952; (b) Neuman, W. P. "The Organic Chemistry of Tin", Wiley: New York, 1970; pp. 230-237; (c) Krigman, M. R.; Silverman, A. P. *Neurotoxicology*, **1984**, *5*, 129.
3. Bottaro, J. C.; Hanson, R. N.; Seitz, D. E. *J. Org. Chem.* **1981**, *46*, 5221.
4. Hayashi, K.; Iyoda, J.; Shiihara, I. *J. Organomet. Chem.* **1967**, *10*, 81.
5. Corey, E. J.; Wollenberg, R. H. *J. Org. Chem.* **1975**, *40*, 3788.
6. Coulson, D. R. *Inorg. Synth.* **1972**, *13*, 121.
7. Fitton, P.; McKeon, J. E.; Ream, B. C. *J. Chem. Soc., Chem. Commun.* **1969**, 370.
8. Trost, B. M.; Caldwell, C. G.; Murayama, E.; Heissler, D. *J. Org. Chem.* **1983**, *48*, 3252.

9. (a) Kosugi, M.; Shimizu, Y.; Migita, T. *Chem. Lett.* **1977**, 1423; (b) Kosugi, M.; Shimizu, Y.; Migita, T. *J. Organomet. Chem.*, **1977**, *129*, C36; (c) Milstein, D.; Stille, J. K. *J. Am. Chem. Soc.* **1978**, *100*, 3636; (d) Milstein, D.; Stille, J. K. *J. Org. Chem.* **1979**, *44*, 1613.
10. Sato, F.; Inoue, M.; Oguro, K.; Sato, M. *Tetrahedron Lett.* **1979**, 4303.
11. Posner, G. H. *Org. Reactions* **1975**, *22*, 253.
12. Negishi, E.-I.; Idacavage, M. J. *Org. Reactions* **1985**, *33*, 1.
13. (a) Labadie, J. W.; Stille, J. K. J. Am. Chem. Soc. **1983**, *105*, 6129; (b) Labadie, J. W.; Tueting, D.; Stille, J. K. *J. Org. Chem.* **1983**, *48*, 4634.
14. Yamamoto, Y.; Yanagi, A. *Chem. Pharm. Bull.* **1982**, *30*, 2003.
15. Logue, M. W.; Teng, K. *J. Org. Chem.* **1982**, *47*, 2549.
16. Soderquist, J. A.; Leong, W. W.-H. *Tetrahedron Lett.* **1983**, *24*, 2361.
17. Kende, A. S.; Roth, B.; Sanfilippo, P. J.; Blacklock, T. J. *J. Am. Chem. Soc.* **1982**, *104*, 5808.
18. (a) Corey, E. J.; Wollenberg, R. H. *J. Am. Chem. Soc.* **1974**, *96*, 5581; (b) Nesmeyanov, A. N.; Borisov, A. E. *Dokl. Akad. Nauk SSSR* **1967**, *174*, 96; *Chem. Abstr.* **1967**, *67*, 90903g.
19. Seyferth, D.; Vick, S. C. *J. Organomet. Chem.* **1978**, *144*, 1.
20. Leusink, A. J.; Budding, H. A.; Marsman, J. W. *J. Organomet. Chem.* **1967**, *9*, 285, 295; Leusink, A. J.; Budding, H. A.; Drenth, W. *J. Organomet. Chem.* **1967**, *9*, 295.
21. (a) Piers, E.; Chong, J. M.; Morton, H. E. *Tetrahedron Lett.* **1981**, *22*, 4905; (b) Piers, E.; Chong, J. M. *J. Org. Chem.* **1982**, *47*, 1602.
22. Seitz, D. E.; Lee, S.-H. *Tetrahedron Lett.* **1981**, *22*, 4909.
23. Leibner, J. E.; Jacobus, J. *J. Org. Chem.* **1979**, *44*, 449.

Appendix

Chemical Abstracts Nomenclature (Collective Index Number);

(Registry Number)

Tributylethynylstannane: Stannane, tributylethynyl- (8,9); (994-89-8)

Lithium acetylide, ethylenediamine complex: Ethylenediamine, compd. with lithium acetylide (Li(HC$_2$)) (1:1) (8); 1,2-Ethanediamine, compd. with lithium acetylide (Li(C$_2$H)) (1:1) (9); (6867-30-7)

Tributyltin chloride: Stannane, tributylchloro- (8,9); (1461-22-9)

(E)-1,2-Bis(tributylstannyl)ethylene: Stannane, vinylenebis[tributyl-, (E)- (8); Stannane, 1,2-ethenediylbis[dibutyl-, (E)- (9); (14275-61-7)

p-Nitrobenzoyl chloride: Benzoyl chloride, p-nitro- (8); Benzoyl chloride, 4-nitro- (9); (122-04-3)

Benzylchlorobis(triphenylphosphine)palladium(II);

Palladium, benzylchlorobis(triphenylphosphine)-, trans- (8);

Palladium, chloro(phenylmethyl)bis(triphenylphosphine)-,

(SP-4-3)- (9); (22784-59-4)

Tetrakis(triphenylphosphine)palladium(0); Palladium, tetrakis(triphenylphosphine)- (8); Palladium, tetrakis(triphenylphosphine)-, (T-4)- (9); (142210-01-3)

Pyridinium poly(hydrogen fluoride): Pyridine, compd. with hydrofluoric acid (1:1); Pyridine, hydrofluoride (9); (32001-55-1)

ETHYL 5-OXO-6-METHYL-6-HEPTENOATE FROM METHACRYLOYL CHLORIDE AND ETHYL 4-IODOBUTYRATE

Submitted by Yoshinao Tamaru, Hirofumi Ochiai, Tatsuya Nakamura, and Zen-ichi Yoshida.[1]
Checked by Kevin B. Kunnen and Albert I. Meyers.

1. Procedure

A 300-mL, four-necked, round-bottomed flask containing a magnetic stirring bar is fitted with a serum-cap, thermometer, 100-mL serum-capped pressure-equalizing addition funnel, and a reflux condenser equipped at the top with a nitrogen inlet. The dry apparatus is flushed with nitrogen and 5.6 g (85.5 mmol) of zinc-copper couple (Note 1) and 20 mL of benzene are introduced (Note 2). A mixture of 13.8 g (57 mmol) of ethyl 4-iodobutyrate (Note 3), 9 mL of N,N-dimethylacetamide (Note 4), and 70 mL of benzene is transferred into the addition funnel by cannulation techniques and added to the stirred Zn(Cu) suspension over 3 min at room temperature. The mixture is vigorously stirred for 1 hr at room temperature (Note 5) and then heated at gentle reflux with an oil bath for 4.5 hr (Note 6). After the mixture is cooled to 60°C, a solution of 0.58 g (0.5 mmol) of tetrakis(triphenyl-phosphine)palladium(0) (Note 7) in 15 mL of benzene is added over 1 min through the addition funnel and stirring is continued for 5 min at the same temperature. The oil bath is removed, a solution of 5.23 g (50 mmol) of methacryloyl chloride (Note 8) in 10 mL of benzene is added through the addition funnel over a period of 5 min, and stirring is continued for 1 hr (Note 9). The mixture is filtered with suction through a Celite pad on a medium-fritted funnel and the filter cake is washed with 200 mL of diethyl ether. The filtrate is washed successively with 50 mL of 1 N ammonium chloride, 10 mL of saturated sodium hydrogen carbonate and 50 mL of saturated sodium chloride. The aqueous phases are extracted with 100 mL of diethyl ether. The combined organic extracts are dried over magnesium sulfate and the solvents are removed with a rotary evaporator to yield a deep brown mobile oil. This is purified by chromatography on 200 g of silica gel with a hexane-diethyl ether gradient (10:1, 400 mL; 5:1, 400 mL, and 2:1, 600 mL) (Note 10), followed by distillation in the presence of hydroquinone (10 mg) in a Kugelrohr apparatus to give 8.0-8.1 g (87-88%) of the product as a colorless liquid, bp 185°C (20 mm) (Note 11).

2. Notes

1. Zinc-copper couple was prepared according to the literature procedure[2] and kept in a desiccator over phosphorus pentoxide under nitrogen.

2. Benzene is dried by distillation from sodium/benzophenone ketyl.

3. Ethyl 4-iodobutyrate, bp 65°C (2.5 mm), was obtained according to the literature procedure[3] (80-90% yield). A mixture of 50 g (0.26 mol) of ethyl 4-bromobutyrate, available from Aldrich Chemical Company, Inc, and 190 g (1.26 mol) of sodium iodide was heated in 500 mL of acetone at 60°C for 24 hr.

4. N,N-Dimethylacetamide (DMA) is dried by distillation under reduced pressure from calcium hydride. The use of DMA is essential to promote metallation.[4] The metallation is also successful with N,N-dimethylformamide as a co-solvent, but the yield of product is significantly lowered (60-70%) because of formation of acid anhydride.[4b]

5. The metallation is only slightly exothermic.

6. It is difficult to judge the completion of the metallation by appearance. Although the metallation is reproducible, it is recommended that a reaction aliquot be checked by VPC or TLC after quenching with 1 N hydrochloric acid.

7. Tetrakis(triphenylphosphine)palladium(0) is available from Aldrich Chemical Company, Inc.

8. Methacryloyl chloride, obtained from Aldrich Chemical Company, Inc., is distilled from calcium chloride under nitrogen at atmospheric pressure into a flask containing a small amount of hydroquinone monomethyl ether.

9. The reaction is moderately exothermic and the temperature rises to about 65°C after the addition of methacryloyl chloride and then gradually falls to ambient temperature.

10. The reaction mixture gives only one spot on silica gel TLC (R_f = 0.5, hexane/ethyl acetate = 4:1 using iodine or saturated 2,4-dinitrophenylhydrazine in 2 N hydrochloric acid as an indicator). This column purification is undertaken to help the smooth distillation of the product. Silica gel 60 MERCK in a 5.5-cm diameter column was used.

11. The submitters report bp 100°C (6 mm). The product shows the correct elemental analysis and has the following physical and spectral properties: n_D^{20} 1.4512; IR (liquid film) cm^{-1}: 3100 (w), 2970 (m), 1730 (s), 1680 (s), 1630 (m), 940 (m); ^{1}H NMR (CDCl$_3$) δ: 1.25 (t, 3 H, J = 7.1), 1.75-2.11 (m, 5 H), 2.35 (t, 2 H, J = 6.8), 2.76 (t, 2 H, J = 7.1), 4.13 (q, 2 H, J = 7.1), 5.77 (br, s, 1 H), 5.96 (s, 1 H); ^{13}C NMR (CDCl$_3$) δ: 13.9, 17.2, 19.3, 33.0, 36.0, 59.9, 124.1, 144.1, 172.7, 200.6; VPC analysis: 20% Silicone DC550 on Celite (Nishio Kogyo Co.), 3 mm x 1 m column, constant temperature increase 8°C/min from 100°C, one peak of impurity (retention time 1.8 min) and the peak of the product (retention time 9.0 min, 99.5% purity).

3. Discussion

α-Metallocarbonyl compounds, so called "enolates", are among the most widely used reagents for organic syntheses. Undoubtedly, β-, γ-, and δ-metallocarbonyl compounds are also of great synthetic value. The use of these organometallics, however, has been limited mainly because of the lack of a convenient preparative method. Herein are described the preparation of a γ-metallo ester, 3-carboethoxypropylzinc iodide (**1**, n = 3; eq 1), and its reaction with acid chlorides to yield δ-keto esters. According to the same procedure, it is possible to generate 2-carboethoxyethylzinc iodide (**1**, n = 2) and 4-carboethoxybutylzinc iodide (**1**, n = 4)[5] with similar efficiency. 2-Carboalkoxyethylzinc may be prepared by a cyclopropane ring-opening procedure (eq 2).[6]

$$EtO_2C(CH_2)_nI + Zn(Cu) \longrightarrow EtO_2C(CH_2)_nZnI \qquad (1)$$
$$\mathbf{1}$$

$$2 \; \triangleright\!\!<^{OSiMe_3}_{OR} \; + \; ZnCl_2 \; \longrightarrow \; (RO_2CCH_2CH_2)_2Zn \; + \; 2 \; Me_3SiCl \quad (2)$$

These organozincs, **1** (n = 2, 3, and 4), react with diverse acid chlorides to yield γ, δ, and ε-keto esters, respectively, in good yields. Two typical examples are shown in eq 3 and 4. The product of eq 3, ethyl 5-oxo-6-heptenoate, may be prepared by laborious, multi-step methods.[7] The title compound, ethyl 5-oxo-6-methyl-6-heptenoate, is a new compound.

1 (n = 3) + $CH_2=CHCOCl$ $\xrightarrow{89\%}$ [structure: ethyl 5-oxo-6-heptenoate] (3)

1 (n = 3) + $MeO_2C(CH_2)_4COCl$ $\xrightarrow{90\%}$ [structure: keto diester] (4)

Another use of **1** (n = 3) is the coupling with vinyl iodides or triflates, which furnish δ,ε-unsaturated esters.[8] One example is shown in eq 5. The reaction proceeds with retention of the double bond geometry.

1 (n = 3) + n-Bu-CH=CH-I $\xrightarrow{89\%}$ n-Bu-CH=CH-CH$_2$CH$_2$-CO$_2$Et (5)

1. Department of Synthetic Chemistry, Faculty of Engineering, Kyoto University, Yoshida, Sakyo, Kyoto 606, Japan.
2. Smith, R D.; Simmons, H. E. *Org. Synth., Collect. Vol. 5*, **1973**, 855.
3. Fuson, R. C.; Arnold, R. T.; Cooke, H. G. Jr. *J. Am. Chem. Soc.* **1938**, *60*, 2272.
4. (a) Tamaru, Y.; Ochiai, H.; Sanda, F.; Yoshida, Z. *Tetrahedron Lett.* **1985**, *26*, 5529; (b) Tamaru, Y.; Ochiai, H.; Nakamura, T.; Tsubaki, K.; Yoshida, Z. *Tetrahedron Lett.* **1985**, *26*, 5559.
5. Tamaru, Y.; Ochiai, H.; Nakamura, T.; Yoshida, Z., unpublished results.
6. (a) Nakamura, E.; Kuwajima, I. *J. Am. Chem. Soc.* **1984**, *106*, 3368; (b) Nakamura, E.; Kuwajima, I. *Org. Synth.* to be published.
7. (a) Barkley, L. B.; Knowles, W. S.; Raffelson, H.; Thompson, Q. E. *J. Am. Chem. Soc.* **1956**, *78*, 4111; (b) Vig, O. P.; Dhindsa, A. S.; Vig, A. K.; Chugh, O. P. *J. Indian Chem. Soc.* **1972**, *49*, 163.
8. Tamaru, Y.; Ochiai, H.; Nakamura, T.; Yoshida, Z. *Tetrahedron Lett.* **1986**, *27*, 955.

Appendix

Chemical Abstracts Nomenclature (Collective Index Number); (Registry Number)

Zinc-copper couple: Copper, compd. with zinc (1:1) (8,9); (12019-27-1)

Ethyl 4-iodobutyrate: Butyric acid, 4-iodo-, ethyl ester (8); Butanoic acid, 4-iodo-, ethyl ester (9); (7425-53-8)

Ethyl 4-bromobutyrate: Butyric acid, 4-bromo-, ethyl ester (8); Butanoic acid, 4-bromo-, ethyl ester (9); (2969-81-5)

N,N-Dimethylacetamide: Acetamide, N,N-dimethyl- (8,9); (127-19-5)

Tetrakis(triphenylphosphine)palladium(0): Palladium, tetrakis(triphenylphosphine)- (8); Palladium, tetrakis(triphenylphosphine)-, (I-4)- (9); (14221-01-3)

Methacryloyl chloride (8); 2-Propenoyl chloride, 2-methyl- (9); (920-46-7)

Hydroquinone (8); 1,4-Benzenediol (9); (123-31-9)

1,4-FUNCTIONALIZATION OF 1,3-DIENES VIA PALLADIUM-CATALYZED CHLOROACETOXYLATION AND ALLYLIC AMINATION: 1-ACETOXY-4-DIETHYLAMINO-2-BUTENE AND 1-ACETOXY-4-BENZYLAMINO-2-BUTENE

(2-Buten-1-ol, 4-(diethylamino)-, acetate (ester))

A. $\diagup\diagdown\diagup$ + LiCl + LiOAc · $2H_2O$ $\xrightarrow[\text{HOAc, 25°C}]{\text{7.5 mol\% Pd(OAc)}_2 \\ \text{p-benzoquinone}}$ AcO$\diagdown\diagup\diagdown$Cl

B. AcO$\diagdown\diagup\diagdown$Cl + RR'NH $\xrightarrow[\substack{\text{or 80°C} \\ \text{uncatalyzed} \\ \text{in CH}_3\text{CN}}]{\text{4 mol\% Pd(PPh}_3)_4 \\ \text{25°C, THF}}$ AcO$\diagdown\diagup\diagdown$NRR'

R = R' = Et
R = H, R' = $PhCH_2$

Submitted by J. E. Nyström, T. Rein, and J. E. Bäckvall.[1]
Checked by Marvin M. Hansen and Clayton H. Heathcock.

1. Procedure

A. *1-Acetoxy-4-chloro-2-butene.* In a 2-L, two-necked, round-bottomed flask equipped with a 5-cm egg-shaped magnetic stirring bar and a pressure-reducing outlet (Note 1) is placed 800 mL of pentane (Note 2). The flask is cooled with an ice bath and 5.4 g (0.1 mol) of butadiene is dissolved with stirring (0-5°C) by addition through one of the inlets from a Fluka low pressure bottle of butadiene (Note 3). The pressure-reducing outlet is removed and a freshly prepared solution of 1.68 g (7.5 mmol) of palladium acetate, $Pd(OAc)_2$, 8.4 g (0.2 mol) of lithium chloride, 20.4 g (0.2 mol) of lithium acetate dihydrate (LiOAc $2H_2O$), and 21.6 g (0.2 mol) of p-benzoquinone in 400 mL of acetic acid is added (Note 4). The cooling bath is removed and the two-phase system is stirred at a moderate rate (Note 5) at 25°C for 26 hr. A saturated sodium chloride solution (300 mL) is added and after the mixture is stirred for 5 min, it is filtered using a Büchner funnel with an intermediate paper filter using aspirator vacuum. The organic phase is separated and the aqueous phase is extracted with three 300-mL portions of pentane-ether (80:20). The combined organic phases are washed with two 75-mL portions of water, two 100-mL portions of saturated potassium carbonate solution, three 100-mL portions of 2M sodium hydroxide solution, and finally with 50 mL of saturated sodium chloride solution. The organic phase is dried over anhydrous magnesium sulfate, filtered and concentrated to a volume of 20-30 mL by distilling off the solvent at atmospheric pressure. The remaining solvent is removed with a rotary evaporator at aspirator vacuum to give 13-15 g of crude product, which is distilled (10 mm, 70-90°C) to yield 9.7-12.0 g (65-81%) of a light yellow liquid consisting of 91% of 1-acetoxy-4-chloro-2-butene (E/Z = 90/10) and 9% of 4-acetoxy-3-chloro-1-butene. The chloroacetate is contaminated with approximately 1% of 5,8-dihydronaphthoquinone (Note 6).

Further purification is achieved by the following procedure: The chloroacetate from above is dissolved in 150 mL of ether. This solution is stirred together with a 10-mL aqueous solution saturated with sodium borohydride. The stirring is continued until the yellow color of the organic phase disappears (ca. 15 min). The organic phase is separated and washed with 5 mL of 2M sodium hydroxide solution, 5 mL of a saturated sodium chloride solution, dried over magnesium sulfate and concentrated by distilling off the solvent at atmospheric pressure. The solvent which remains is removed with a rotary evaporator to afford 9.5-10.5 g (64-70%) of chloroacetate, with the same composition as above, but which is now completely free from 5,8-dihydronaphthoquinone (Note 7).

B1. 1-Acetoxy-4-diethylamino-2-butene. Method 1. In a 500-mL, two-necked, round-bottomed flask equipped with a magnetic stirring bar, nitrogen-vacuum inlet, and a rubber septum, is placed 2.77 g (2.4 mmol) of tetrakis(triphenylphosphine)palladium, $Pd(PPh_3)_4$ (Note 8). The flask is closed, evacuated and filled with nitrogen. This flushing procedure is repeated twice (Note 9). A solution of 8.91 g (0.06 mol) of the chloroacetate from procedure A in 180 mL of dry tetrahydrofuran (Note 10) is added through the membrane with the aid of a 50-mL syringe. With the same syringe 21.9 g (0.30 mol) of diethylamine (Note 11) in 120 mL of dry tetrahydrofuran is added. The mixture is stirred at ambient temperature and the reaction is followed by gas chromatography. When the starting material has been consumed, which takes approximately 4 hr (Note 12), 600 mL of ice-cooled ether and 600 mL of ice-cooled saturated aqueous sodium carbonate solution are added and the mixture is shaken in a separatory funnel (Note 13). The aqueous phase is extracted with ether (2 x 300 mL).

The combined organic phases are washed with 50 mL of saturated potassium carbonate solution, and dried over solid potassium carbonate. Evaporation with a rotary evaporator affords 13.7 g of crude product. The residue is put on a column (silica, 3 x 10 cm) and eluted with 600 mL of ether-pentane-triethylamine (47.5:47.5:5) (Note 14). The main fractions are collected to give 8.15 g (73%) of essentially pure E-1-acetoxy-4-diethylamino-2-butene (>94% E). No 1,2-isomer could be detected. The product is further purified by Kugelrohr distillation to afford 7.81-8.21 g (70-74%) (Note 15).

B2. 1-Acetoxy-4-benzylamino-2-butene. Method 2. In a 250-mL, one-necked, round-bottomed flask are placed in order 8.91 g (0.06 mol) of the chloroacetate from method A, 100 mL of acetonitrile and 19.3 g (0.18 mol) of benzylamine (Note 6). The flask is equipped with a reflux condenser and the solution is refluxed for 2 hr using an oil bath at 100-110°C. The reaction mixture is cooled and 150 mL of ether and 100 mL of a saturated sodium carbonate solution are added. The mixture is shaken in a separatory funnel and the organic phase is collected. The aqueous phase is extracted with 50 mL of ether. The combined organic phases are dried over potassium carbonate. The solvent and excess benzylamine are removed by rotary evaporation and Kugelrohr distillation at 50°C (1 mm). Kugelrohr distillation of the crude product gives 9.1-9.9 g (70-76%) of 1-acetoxy-4-benzylamino-2-butene as a 90:10 mixture of the E and Z isomers (Note 17).

2. Notes

1. The pressure-reducing outlet can be a U-shaped tube filled with oil or a thick-walled rubber balloon.

2. Light petroleum, boiling point 40°C, can also be used.

3. The amount of butadiene added is determined by weighing the Fluka bottle and double checked by weighing the reaction flask before and after addition. The checkers purchased butadiene from Matheson and measured it by condensation into a 25-mL flask cooled to -10°C. The cooled material was then transferred by cannula into the reaction vessel containing the pentane, cooled to 0-5°C.

4. p-Benzoquinone, 200 mol%, is needed for a rapid and efficient reaction.

5. A stirring rate of approximately 5 rps is used. The reaction tolerates a variation between 3-10 rps, which gives essentially the same result.

6. 5,8-Dihydronaphthoquinone is the oxidized Diels-Alder adduct between butadiene and benzoquinone.

7. The spectral properties are as follows: ^{1}H NMR (CDCl$_3$, 250 MHz) δ: 2.09 (s, 3 H), 4.06 (m, 2 H), 4.59 (m, 2 H), 5.90 (m, 2 H).

8. Tetrakis(triphenylphosphine)palladium, Pd(PPh$_3$)$_4$, is commercially available but is readily prepared according to ref 2 (or ref 3b). Palladium acetylacetonate, Pd(acac)$_2$, together with 4 PPh$_3$ can be used in place of Pd(PPh$_3$)$_4$ and gives essentially the same result.

9. A manifold system connected to a vacuum line and a nitrogen line is used.

10. Tetrahydrofuran is distilled under nitrogen from potassium benzophenone.

11. Commercial diethylamine (BDH) is used without further purification.

12. The rate of the reaction varies slightly depending on the quality of the catalyst.

13. The solution is kept cold to avoid hydrolysis of the acetoxy group.

14. The checkers observed that, upon placing the crude product on the top of the silica gel column, the residual Pd(PPh$_3$)$_4$ precipitates. However, the presence of this solid residue does not interfere with the progress of the chromatography or affect the yield of product.

15. The spectral properties are as follows: ^{1}H NMR (CDCl$_3$, 250 MHz) δ: 1.03 (t, 6 H, J = 7.2), 2.07 (s, 3 H), 2.51 (q, 4 H, J = 7.2), 3.10 (br d, 2 H), 4.55 (br d, 2 H), 5.64-5.90 (m, 2 H).

16. Benzylamine (Fluka) is dried over NaOH and distilled over sodium. The threefold excess of benzylamine is used to depress the dialkylation product.

17. The spectral properties are as follows: ^{1}H NMR (CDCl$_3$, 250 MHz) δ: 1.20-1.60 (br s, 1 H), 2.06 (s, 3 H), 3.28 (br d, 2 H, J = 5.7), 3.78 (s, 2 H), 4.55 (br d, 2 H, J = 5.8), 5.74 (dt, 1 H, J = 15.5, 5.8), 5.88 (dt, 1 H, J = 15.7, 5.7), 7.20-7.45 (m, 5 H).

3. Discussion

The procedure reported here provides an efficient method for the preparation of 4-amino-2-alken-1-ol derivatives. It is based on the palladium-catalyzed 1,4-acetoxychlorination of 1,3-dienes[3] and palladium-catalyzed amination of allylic substrates.[4] Compared to other methods[5] this method is more convenient and more general. It allows complete control of the 1,4-relative configuration when the carbons bearing nitrogen and oxygen are stereogenic. In these cases the chloride is replaced with *retention* according to procedure B1 but with *inversion* according to procedure B2.[3b,6]

Procedure A is very effective for a range of acyclic and cyclic conjugated dienes.[3] The major side reaction in the chloroacetoxylation is Diels-Alder addition of p-benzoquinone to the diene. The purpose of the pentane phase is to ensure a low concentration of diene in the acetic acid phase, which depresses the Diels-Alder reaction. The reaction can also be performed without the pentane phase with slow addition of the diene using a syringe pump.

Some representative examples of the amination reaction according to procedure B are shown in Table I.

1. Department of Organic Chemistry, Royal Institute of Technology, S-100 44 Stockholm, Sweden.
2. Coulson, D. R. *Inorg. Synth.* **1972**, *13*, 121.
3. (a) Bäckvall, J. E.; Nordberg, R. E.; Nyström, J. E. *Tetrahedron Lett.* **1982**, *23*, 1617; (b) Bäckvall, J. E.; Nyström, J. E.; Nordberg, R. E. *J. Am. Chem. Soc.* **1985**, *107*, 3676.
4. Genêt, J. P.; Balabane, M.; Backvall, J. E.; Nyström, J. E. *Tetrahedron Lett.* **1983**, *24*, 2745.
5. Tsuji, J.; Kataoka, H.; Kobayashi, Y. *Tetrahedron Lett.* **1981**, *22*, 2575.
6. Backvall, J. E. *Pure Appl. Chem.* **1983**, *55*, 1669.

Appendix
Chemical Abstracts Nomenclature (Collective Index Number);
(Registry Number)

1-Acetoxy-4-diethylamino-2-butene: 2-Buten-1-ol, 4-(diethylamino)-, acetate (ester) (11); (82736-47-8)

1-Acetoxy-4-chloro-2-butene: 2-Buten-1-ol, 4-chloro-, acetate (E)- (9); (34414-28-3)

Palladium acetate: Acetic acid, palladium(2+) salt (8,9); (3375-31-3)

Lithium acetate dihydrate: Acetic acid, lithium salt, dihydrate (8,9); (6108-17-4)

4-Acetoxy-3-chloro-1-butene: 3-Buten-1-ol, 2-chloro-, acetate (11); (96039-67-7)

Tetrakis(triphenylphosphine)palladium:

Palladium, tetrakis(triphenylphosphine)- (8);

Palladium, tetrakis(triphenylphosphine)-, (T-4)- (9); (14221-01-3)

TABLE

Chloroacetate[a]	Amine	Procedure	Aminoacetate	Yield(%)
AcO-CH=CH-CH₂Cl	Et₂NH	B1	AcO-CH=CH-CH₂-NEt₂	87[b]
		B2		75
AcO-C(Me)=CH-CH₂Cl[c]	EtNH₂	B2	AcO-C(Me)=CH-CH₂-NHEt[c]	89
	Me₂NH	B2	AcO-C(Me)=CH-CH₂-NMe₂[c]	76
	PhCH₂NH₂	B2	AcO-C(Me)=CH-CH₂-NHCH₂Ph[c]	89
(OAc)CH-CH=CH-CH(Cl)[d]	Et₂NH	B1	(OAc)CH-CH=CH-CH(NEt₂)[d]	80
		B2	(OAc)CH-CH=CH-CH(NEt₂)[e]	70
cyclohexenyl-OAc/Cl[f]	Me₂NH	B1	cyclohexenyl-OAc/NMe₂[f]	95
		B2	cyclohexenyl-OAc/NMe₂[g]	93
cycloheptenyl-(OAc)₂/Cl[h]	MeNH₂	B1	cycloheptenyl-(OAc)₂/NHMe[h]	87

a) The chloroacetates were prepared from the corresponding dienes according to procedure A or the modified version without the pentane phase (see discussion). b) From a small scale experiment with acid extraction in the work-up. c) E/Z = 3.5/1. d) >94% R*R*. e) >90% R*S*. f) >98% cis. g) >98% trans. h) >95% β,β,α.

PALLADIUM(0)-CATALYZED syn-1,4-ADDITION OF CARBOXYLIC ACIDS TO CYCLOPENTADIENE MONOEPOXIDE: cis-3-ACETOXY-5-HYDROXYCYCLOPENT-1-ENE

(4-Cyclopentene-1,3-diol, monoacetate, cis-)

Submitted by Donald R. Deardorff and David C. Myles.[1]
Checked by Helmut Grebe and Ekkehard Winterfeldt.

1. Procedure

An oven-dried, 300-mL, two-necked, round-bottomed flask is equipped with a Teflon-coated magnetic stirring bar, pressure equalizing dropping funnel, and a rubber septum with an 18-gauge needle connected to a dry nitrogen source. The nitrogen-flushed apparatus is charged with 125 mL of dry tetrahydrofuran (Note 1) and 0.28 g (0.24 mmol, 0.2 mol%) of tetrakis(triphenylphosphine)palladium(0) (Note 2). The mixture is stirred at room temperature until all of the palladium catalyst dissolves (Note 3). The solution is cooled in an ice-water bath and 7.0 mL (7.3 g, 122 mmol) of acetic acid (Note 4) is added via syringe. At this point a slight darkening of the solution is observed. A room temperature solution containing 10.9 g of 92% cyclopentadiene monoepoxide (10.0 g, 122 mmol, Note 5) in 40 mL of tetrahydrofuran is added over 10 min with the aid of the addition funnel. The original pale yellow color gives way to a deeper transparent orange. After 5 min (Note 6), the solution is concentrated at ambient temperature under reduced pressure and the resulting reddish-brown oil is passed through a plug of silica gel (50 g, Note 7) with 450 mL of ethyl ether (Note 8). The slightly cloudy filtrate is washed through a plug of anhydrous magnesium sulfate (60 g, 5 x 7 cm coarse glass frit) with an additional 150 mL of ether (Note 9). The solvent is removed under reduced pressure to yield a pale yellow oil. The material is distilled through a short path apparatus at 73-75°C (0.15 mm) to afford 12.5-13.2 g (72-76%) of colorless oil which crystallizes upon refrigeration (mp 36-39°C). The material is homogeneous by TLC and ^{1}H NMR (Note 10), but can be further purified by recrystallization from an ether-hexane mixture to give colorless crystals, mp 38.5-41°C, (Note 11).

2. Notes

1. Tetrahydrofuran was predried over potassium hydroxide (*Caution!* See *Org. Synth., Collect. Vol. V* **1973**, 976 *for possible hazard*.), then dried by distillation from sodium/benzophenone ketyl under nitrogen.

2. Tetrakis(triphenylphosphine)palladium(0) was purchased from Aldrich Chemical Company Inc., and used without further purification. No special precautions were taken in handling the catalyst.

3. Dissolution takes approximately 2-3 min.

4. Glacial acetic acid was purchased from J. T. Baker Chemical Company (Baker Analyzed Reagent) and distilled prior to use.

5. Cyclopentadiene monoepoxide was prepared from cyclopentadiene and peracetic acid according to the well-established procedure of Korach et al.[2] However, the submitters report that epoxidation of cyclopentadiene under conditions developed by Knapp et al.[3] for cyclohexadiene increased their isolated yields from 40% to 62%. The major impurity in distilled

cyclopentadiene monoepoxide is 3-cyclopentenone. This by-product can be conveniently assayed by ^{1}H NMR integration of the four methylene protons which appear as a broad singlet at 2.80 δ. Epoxide purity can be determined by GLC analysis on a 25 M Carbowax capillary column operated at 50°C.

6. TLC analysis using Baker Si250F precoated glass plates with a hexane-ethyl acetate (1:1) solvent system indicates that all of the starting material is consumed.

7. Baker Analyzed Reagent silica gel 60-200 mesh was used in a 4.5 x 9.0-cm column. This step removes most palladium-containing compounds from the reaction mixture. In order to insure that the palladium is efficiently separated, it is important that all tetrahydrofuran be removed from the crude oil prior to filtration.

8. Anhydrous ether (purified) was purchased from J. T. Baker Chemical Company and used without additional purification.

9. This step removes the final traces of palladium. It is imperative that all catalyst be removed prior to distillation since upon heating this metal catalyzes the decomposition of cis-3-acetoxy-5-hydroxycyclopent-1-ene into isomeric cyclopentenones and acetic acid.

10. cis-3-Acetoxy-5-hydroxycyclopent-1-ene has the following spectral characteristics: ^{1}H NMR (200 MHz, CDCl$_3$) δ: 1.59 (dt, 1 H, J = 4.0 and 14.5, CH$_2$), 2.00 (s, 3 H, CH$_3$), 2.38 (br s, 1 H, OH), 2.76 (quintet, overlapping dt, 1 H, J = 7.4 and 14.5, CH$_2$), 4.67 (m, 1 H, CHOH), 5.45 (m, 1 H, CHOAc), 5.92 (m, 1 H, CH=CH), 6.05 (m, 1 H, CH=CH); IR (neat) cm^{-1}: 3410 (s), 1720 (s), 1250 (s).

11. 4-Acetoxy-2-cyclopentenone could be prepared in 85% yield by treatment of 1 equiv of this alcohol with 1.1 equiv of pyridinium chlorochromate in methylene chloride for 1 hr at room temperature, followed by washing with water, concentration, and distillation.

3. Discussion

Functionalized cyclopentenoids have been used extensively as key building blocks for the synthesis of many biologically active molecules.[4] This procedure details the facile preparation of one such versatile intermediate: cis-3-acetoxy-5-hydroxycyclopent-1-ene. Although important in its own right, this material also serves as a one-step precursor to the highly useful synthetic substrate 4-acetoxy-2-cyclopenten-1-one.[4] Only the acetic acid adduct with cyclopentadiene monoepoxide is described here. However, this palladium-catalyzed reaction appears general for other acidic substrates as well.[5,6] For example, the corresponding benzoate and phenyl ether adducts have been successfully prepared[5] from both benzoic acid and phenol in yields of 87% and 82%, respectively. Moreover, the reaction is not limited to just the monoprotected versions of cis-cyclopentene-3,5-diol. The corresponding diesters can be similarly prepared by replacement of the carboxylic acid component with an anhydride. This minor modification permits direct synthetic access to either the dibenzoate or diacetate in equally good yields (74% and 79% respectively). Recently, silyl carboxylates and silyl phenoxides were also found to react analogously with cyclopentadiene monoepoxide in the presence of Pd(0) catalyst.[7] It should be stressed that in each case only the cis-1,4-adducts are observed, despite the fact that three other stereoisomers are possible. This remarkable stereo- and regiospecificity is undoubtedly a manifestation of an intermediate palladium π-allyl complex.[8]

Racemic cis-monoesters of cyclopentene-3,5-diol have been previously prepared by the selective acylation[9] of the meso-diol and the copper-mediated[10] addition of carboxylic acid salts to cyclopentadiene monoepoxide. Optically active monoacetates can be accessed by enzymatic hydrolysis[11] of the corresponding diester. The present method offers four principal advantages over the earlier reports: (1) it is operationally simple, (2) it requires a much shorter reaction time, (3) it gives better yields, and (4) it has widespread applicability, since reactants other than carboxylic acids may be employed with equally good results.

A major disadvantage with the acylation method[9] is that the starting material, cis-cyclopentene-3,5-diol, is not readily available and must be prepared via photoxygenation procedures.[12] Furthermore, acylation occurs with the concomitant formation of diacylated product which results in reduced yields and associated purification problems. The copper-mediated[10] and palladium-catalyzed procedures share some similarities in that they both use cyclopentadiene monoepoxide as their starting material and deliver the desired product in good yield. But, unlike the palladium-catalyzed method, copper-mediated reactions require two full equivalents of carboxylate salt, much lower reaction temperatures (-78°C), and substantially longer reaction times. Finally, the enantioselective hydrolysis[11] of cis-3,5-diacetoxycyclopent-1-ene by hydrolase enzymes is an effective two-step method for generating optically-enriched product.

1. Department of Chemistry, Occidental College, Los Angeles, CA 90041. This work was supported by a Penta Corporation Grant of Research Corporation.
2. Korach, M.; Nielson, D. R.; Rideout, W. H. *Org. Synth., Collect. Vol. 5* **1973**, 414-418.
3. Knapp, S.; Sabastian, M. J.; Ramanathan, H. *J. Org. Chem.* **1983**, *24*, 4786.
4. Harre, M.; Raddatz, P.; Walenta, R.; Winterfeldt, E. *Angew. Chem. Intern. Ed. Engl.* **1982**, *21*, 480, and references therein.
5. Deardorff, D. R.; Myles, D. C.; MacFerrin, K. D. *Tetrahedron Lett.* **1985**, *26*, 5615.
6. Trost, B. M.; Angle, S. R. *J. Am. Chem. Soc.* **1985**, *107*, 6123, ref. 7.
7. Deardorff, D. R.; Shambayati, S.; Linde II, R. G.; Dunn, M. M. *J. Org. Chem.* **1988**, *53*, 189.
8. Keinan, E.; Roth, Z. *J. Org. Chem.* **1983**, *48*, 1769.
9. (a) Tanaka, T.; Kurozumi, S.; Toru, T.; Miura, S.; Kobayashi, M.; Ishimoto, S. *Tetrahedron* **1976**, *32*, 1713; (b) Trost, B. M.; McDougal, P. G. *Tetrahedron Lett.* **1982**, *23*, 5497; (c) Nara, M.; Terashima, S.; Yamada, S. *Tetrahedron* **1980**, *36*, 3161.
10. Marino, J. P.; Jaen, J. C. *Tetrahedron Lett.* **1983**, *24*, 441.
11. (a) Deardorff, D. R.; Matthews, A. J.; McMeekin, D. S.; Craney, C. L. *Tetrahedron Lett.* **1986**, *27*, 1255; (b) Laumen, K.; Reimerdes, E. H.; Schneider, M.; Gorisch, H. *Tetrahedron Lett.* **1985**, *26*, 407; (c) Laumen, K.; Schneider, M. *Tetrahedron Lett.* **1984**, *25*, 5875; (d) Wang, Y-F.; Chen, C-S.; Girdaukas, G.; Sih, C. J. *J. Am. Chem. Soc.* **1984**, *106*, 3695; (e) Miura, S.; Kurozumi, S.; Toru, T.; Tanaka, T.; Kobayashi, M.; Matsubara, S.; Ishimoto, S. *Tetrahedron* **1976**, *32*, 1893.
12. Kaneko, C.; Sugimoto, A.; Tanaka, S. *Synthesis* **1984**, 876.

Appendix

Chemical Abstracts Nomenclature (Collective Index Number);

(Registry Number)

Cyclopentadiene monoepoxide: 6-Oxabicyclo[3.1.0]hex-2-ene (8,9); (7129-41-1)

cis-3-Acetoxy-5-hydroxycyclopent-1-ene: 4-Cyclopenten-1,3-diol, monoacetate, cis- (9); (60410-18-6)

Tetrakis(triphenylphosphine)palladium(0):

Palladium, tetrakis(triphenylphosphine)- (8);

Palladium, tetrakis(triphenylphosphine)-, (I-4)- (9); (14221-01-3)

4,4-DIMETHYL-2-CYCLOPENTEN-1-ONE

(2-Cyclopenten-1-one, 4,4-dimethyl-)

Submitted by David Pauley, Frank Anderson, and Tomas Hudlicky.[1]
Checked by David M. Fink and Andrew S. Kende.

1. Procedure

A. *2,2-Dimethyl-4-oxopentanal* (Note 1). Oxygen is bubbled for 2 hr through a stirred solution of copper(I) chloride (32.8 g, 0.33 mol), palladium(II) chloride (1.19 g, 0.007 mol), 829 mL of dimethylformamide, and 331 mL of water in a 2-L, three-necked, round-bottomed flask cooled in a water bath. 2,2-Dimethyl-4-pentenal (185.6 g, 1.66 mol) (Note 2) is added to the solution and oxygen is bubbled through for an additional 60 hr at room temperature. The solution is acidified to litmus with 10% hydrochloric acid and extracted four times with 200 mL of ethyl ether. The combined organic layers are washed three times with 200 mL of saturated sodium chloride solution and dried over sodium sulfate. The ether is removed first by rotary evaporation and then under reduced pressure (0.2 mm). In this way 133.6 g (1.04 mol) of keto-aldehyde is obtained in 62.7% yield. The original aqueous layer is saturated with sodium chloride and extracted five times with 200 mL of anhydrous ethyl ether. The combined organic layers are washed three times with 200 mL of saturated sodium chloride solution and dried over sodium sulfate. Solvent is removed by rotary evaporation and vacuum pump. An additional 31.5 g (0.25 mol) of keto-aldehyde is recovered to give a total yield of 78% (Note 3); bp 32°C (0.3 mm) (Note 4).

B. *4,4-Dimethyl-2-cyclopenten-1-one.* A 3-L, round-bottomed flask, containing a solution of 600 mL of aqueous 5% potassium hydroxide, 300 mL of tetrahydrofuran, 1350 mL of ethyl ether, and 165.1 g (1.29 mol) 2,2-dimethyl-4-oxopentanal, is equipped with a mechanical stirrer and reflux condenser. The solution is heated under reflux for 66 hr under a nitrogen atmosphere. Upon completion, the organic layer is washed three times with 200 mL of saturated sodium chloride solution and dried over sodium sulfate. The aqueous layer is extracted three times with 200 mL of anhydrous ethyl ether. The resulting organic layers are washed three times with saturated sodium chloride solution and dried over sodium sulfate. All organic layers are evaporated using aspirator vacuum and a rotary evaporator and combined. The residual ether is removed under reduced pressure to yield 89.7 g (63%) of 4,4-dimethyl-2-cyclopenten-1-one, bp 32°C (0.3 mm) (Notes 5-7).

2. Notes

1. This procedure was originally described by Magnus.[2]

2. 2,2-Dimethyl-4-pentenal was prepared as described in *Org. Synth.* **1984**, *62*, 125.

3. The product is sufficiently pure to be used in the next reaction without purification.

4. The spectral properties of the product are as follows: IR (neat) cm^{-1}: 2990, 2730, 1740, 1485, 1380; ^{1}H NMR (CDCl$_3$, 300 MHz) δ: 1.12 (s, 6 H), 2.14 (s, 3 H), 2.7 (s, 2 H), 9.52 (s, 1 H).

5. This product is volatile. The checkers found one-third of the product in a vacuum trap after 24 hr at ca. 5 mm.

6. The checkers distilled the product (Kugelrohr 8 mm, 80°C) and obtained yields of 83% (one-third scale) and 79% (two-thirds scale).

7. The spectral properties of the product are as follows: IR (neat) cm^{-1}: 2970, 2890, 1730, 1600, 1480, 1430; ^1H NMR (CDCl$_3$, 300 MHz) δ: 1.20 (s, 6 H), 2.21 (s, 2 H), 5.97 (d, 1 H, J = 6), 7.46 (d, 1 H, J = 6).

3. Discussion

4,4-Dimethyl-2-cyclopenten-1-one is a valuable starting material in terpenoid synthesis and in cases where a gem-dimethylcyclopentane unit needs to be introduced. It is useful as a starting material in further functionalization. Its preparation by the method of Magnus[2] is amenable to large scale synthesis.

Acknowledgment

The authors thank Professor P. D. Magnus for permission to use his method for the scale up reported here.

1. Department of Chemistry, Virginia Polytechnic Institute and State University, Blacksburg, VA 24061.
2. Magnus, P. D.; Nobbs, M. S. *Synth. Commun.* **1980**, *10*, 273.

Appendix

Chemical Abstracts Nomenclature (Collective Index Number);

(Registry Number)

4,4-Dimethyl-2-cyclopenten-1-one: 2-Cyclopenten-1-one, 4,4-dimethyl- (8,9); (22748-16-9)

2,2-Dimethyl-4-oxopentanal: Pentanal, 2,2-dimethyl-4-oxo- (9); (61031-76-3)

α-DIPHENYLMETHYLSILYLATION OF ESTER ENOLATES:

2-METHYL-2-UNDECENE FROM ETHYL DECANOATE

(2-Undecene, 2-methyl-)

A. $\text{n-C}_8\text{H}_{17}\text{CH}_2\text{CO}_2\text{Et}$ $\quad\xrightarrow[\text{2) Ph}_2\text{MeSiCl}]{\text{1) LDA/THF/-78°C}}\quad$ $\text{n-C}_8\text{H}_{17}\text{CHCO}_2\text{Et}$
$\quad\quad\quad\quad\quad\quad\quad\quad\quad\quad\quad\quad\quad\quad\quad\quad\quad\quad\quad\quad\quad\quad\quad\quad\quad\quad\quad\quad\quad\quad\quad$ |
$\quad\quad\quad\quad\quad\quad\quad\quad\quad\quad\quad\quad\quad\quad\quad\quad\quad\quad\quad\quad\quad\quad\quad\quad\quad\quad\quad\quad\quad\quad\quad$ Ph$_2$SiMe

1

B. $\text{n-C}_8\text{H}_{17}\text{CHCO}_2\text{Et}$ $\quad\xrightarrow[\substack{\text{2) MeLi} \\ \text{3) KOC(CH}_3)_3}]{\text{1) MeMgBr/THF}}\quad$ $\text{n-C}_8\text{H}_{17}\text{CH}\!\!=\!\!\text{C(Me)}_2$
|
Ph$_2$SiMe

2

Submitted by Gerald L. Larson, Ingrid Montes de Lopez-Cepero, and Luis Rodriguez Mieles.[1]

Checked by Choon Sup Ra and Leo A. Paquette.

1. Procedure

A. *Ethyl 2-(diphenylmethylsilyl)decanoate.* A 500-mL, three-necked, round-bottomed flask equipped with a magnetic stirrer, nitrogen inlet, 100-mL pressure-equalizing dropping funnel and a no-air stopper is flame-dried under a vigorous flow of nitrogen, cooled under an atmosphere of nitrogen to -78°C with a dry ice-acetone bath and charged with 39.2 mL (52.5 mmol) of a 1.34 M solution of butyllithium in hexane (Note 1). To this solution is added 7.4 mL (5.31 g; 52.5 mmol) of diisopropylamine (Note 2) in 7 mL of tetrahydrofuran (Note 3). The resulting solution is warmed to ambient temperature and held for 30 min. The solution is diluted with 50 mL of dry tetrahydrofuran and cooled again to -78°C. To this solution is added 11.6 mL (10.0 g; 50 mmol) of ethyl decanoate (Note 4) in 45 mL of tetrahydrofuran dropwise over a 30-min period. The mixture is kept at -78°C for 30 min to allow the enolate to form, and then 10.3 mL (11.6 g; 50 mmol) of diphenylmethylchlorosilane (Note 5) in 40 mL of tetrahydrofuran is added over a 5-min period. The reaction mixture is allowed to reach ambient temperature and stir at that temperature for 8 hr. It then is cooled to 0°C, diluted with hexane (150 mL), washed with cold water (2 x 100 mL), dried over magnesium sulfate, filtered and concentrated at reduced pressure (Note 6). The crude product, which is ca. 95% pure (Note 7), is purified by rapid filtration through 50 g of silica gel (Note 8) with 1% ethyl acetate-hexane (Note 9) as eluant. There is obtained 18.4-18.7 g (93-94%) of ethyl 2-(diphenylmethylsilyl)decanoate (Note 10). Similar results are obtained on a larger scale (Note 11).

B. *2-Methyl-2-undecene*. A 1-L, three-necked, round-bottomed flask equipped with a magnetic stirrer, nitrogen inlet, 500-mL pressure-equalizing dropping funnel and no-air stopper is flame-dried under vacuum, cooled to room temperature under an atmosphere of nitrogen and charged with 87 mL (260 mmol) of 3 M methylmagnesium bromide in ether (Note 12). This solution is cooled to 0°C (ice bath) and 52 g (130 mmol) of ethyl 2-(diphenylmethylsilyl)decanoate in 260 mL of tetrahydrofuran (Note 3) is added over an 8-min period. After the addition is complete, the reaction mixture is warmed to room temperature and heated to reflux for 24 hr. The reaction mixture is again cooled to 0°C and 244 mL (390 mmol) of 1.6 M methyllithium in tetrahydrofuran (Note 13) is added over a period of 30 min. After the addition is complete, the reaction mixture is heated to reflux for 24 hr, cooled to 0°C (ice bath) and 29.2 g (260 mmol) of solid potassium tert-butoxide (Note 14) is added in three portions (Note 15). The reaction mixture is heated to reflux for 1 hr, cooled to 0°C, diluted with hexane (100 mL) and hydrolyzed by the dropwise addition of 1 M hydrochloric acid (240 mL), followed by about 150 mL of 3 M hydrochloric acid until a pH of 4 is reached (Note 16). The organic layer is separated and the aqueous layer is extracted with hexane (3 x 100 mL). The combined organic layers are dried over anhydrous magnesium sulfate, filtered, and concentrated under reduced pressure (Note 6) to give 55 g of crude material (Note 17). This material is diluted with 150 mL of dry hexane (Note 18) and applied to a silica gel column (Note 8). The product is obtained by eluting with hexane and collecting 200-mL fractions. This material, which contains small amounts of dimethyldiphenylsilane and diphenylmethylsilanol, is chromatographed under 3-5 psi on a silica gel column (50 x 2.8 cm) eluting with hexane (Note 19) to give 11.3 g (51.8%) of the olefin (Note 20).

2. Notes

1. Butyllithium was purchased from Foote Mineral Company and titrated by the method of Watson and Eastman.[2]

2. Diisopropylamine was purchased from Aldrich Chemical Company, Inc., and distilled from calcium hydride prior to use.

3. Tetrahydrofuran was a gift from Pfizer Pharmaceuticals of Puerto Rico purchased by them from Dupont Company. It was distilled from sodium/benzophenone prior to use.

4. Ethyl decanoate was purchased from Aldrich Chemical Company, Inc., and used without further purification.

5. Diphenylmethylchlorosilane was purchased from Petrarch Systems, Inc., and distilled from calcium hydride (bp 85°C/0.1 mm) prior to use. A 187.5-mmol scale reaction using diphenylmethylchlorosilane purchased from Petrarch Systems and used without purification gave an 89% yield of the α-silyl ester.

6. A high volume house vacuum system was used for this step.

7. The minor impurities are unreacted ethyl decanoate and diphenylmethylsilanol.

8. Chromatographic silica gel, 70-230 mesh, from Matheson-Coleman-Bell was used.

9. Alternatively, the product can be distilled in an Aldrich Kugelrohr apparatus (pot temperature 130-135°C at 0.2 mm) to give slightly lower (80-90%) yields.

10. The physical properties are as follows: n_D^{20} 1.5190; IR (neat) cm^{-1}: 3068, 3045, 2950-2850, 1714, 1589, 1254, 790; ^{1}H NMR (CDCl$_3$, 80 MHz) δ: 0.66 (s, 3 H), 0.95 (t, 3 H, J = 1), 1.21 (brs, 14 H), 2.56 (m, 1 H), 3.86 (m, 2 H), 7.29-7.62 (m, 10 H); ^{13}C NMR (CDCl$_3$) δ: 3.39, -5.57, 14.01, 22.63, 25.02, 27.56, 29.20, 29.35, 30.51, 31.87, 36.39, 59.75, 127.71, 129.50, 129.56, 134.32, 134.64, 134.78, 134.83, 175.02; MS 70 eV m/e (rel. abundance) 398 (10), 397 (19), 396 (33), 353 (21), 351 (20), 319 (27), 298 (39), 297 (75), 284 (23), 227 (33), 199 (30), 198 (43), 197 (100), 195 (30), 183 (26), 181 (27), 121 (35), 105 (39), 93 (20), 73 (24), 69 (21), 55 (36), 53 (16). Anal. Calcd. for $C_{25}H_{36}O_2Si$: C, 75.76, H, 9.09. Found: C, 75.59, H, 9.19.

11. The submitters report that a 187.5-mmol scale reaction gave an 89% yield of product.

12. Methylmagnesium bromide was purchased from Columbia Organic Chemicals as a 3 M solution and used as obtained.

13. Methyllithium was purchased from Aldrich Chemical Company, Inc., and titrated prior to use.[2]

14. Potassium tert-butoxide was purchased from Aldrich Chemical Company, Inc., and used without further purification.

15. *(CAUTION!) Some foaming occurs because of an exothermic reaction.*

16. Litmus paper was used to determine the pH.

17. Gas chromatographic analysis of this material (6' x 1/8" 10% SP-2401 on 100-120 mesh supelcoport; 100-200°C program at 10°C/min; flow rate of 20 psi) showed the presence of ethyl decanoate, 2-undecanone, dimethyldiphenylsilane, and 2-methyl-2-undecanol in addition to the desired olefin. Small amounts of unidentified products were also present.

18. A mixture of hexanes (Mallinkrodt anhydrous) was used. If the crude product is placed directly on the silica gel column the column plugs and the compound does not elute.

19. Attempts to purify the product by spinning band distillation from the crude material gave only about 20% yield.

20. The product is greater than 97% pure by GLC (Note 17). It showed n_D^{20} 1.4360; ^{1}H NMR (CDCl$_3$, 80 MHz) δ: 0.88 (brt, 3 H), 1.28 (brs, 12 H), 1.61 (brs, 3 H), 1.69 (brs, 3 H), 1.93-2.00 (m, 2 H), 5.14 (m, 1 H); ^{13}C NMR (CDCl$_3$) δ: 14.05, 17.62, 22.74, 25.64, 28.14, 29.41, 29.65, 29.99, 31.98, 125.08, 131.00; MS (70 eV) m/e (rel. abundance) 169 (2), 168 (14), 112 (6), 84 (11), 83 (13), 82 (6), 70 (23), 69 (100), 68 (10), 67 (13), 57 (34), 56 (68), 55 (34), 53 (9).

3. Discussion

Compound 1 represents one example of several α-(diphenylmethylsilyl) esters prepared by the method presented herein.[3] Other examples include the α-diphenylmethylsilylated derivatives of ethyl acetate, ethyl propionate, ethyl 10-undecenoate, ethyl palmitate and ethyl stearate, all obtained in greater than 70% yield. Other alcohols, principally methyl, isopropyl, tert-butyl and 1-menthyl, also have been employed in this reaction without marked differences. The reasons as to why the lithium enolates of esters are silylated at the carbon terminus with diphenylmethylchlorosilane as opposed to the usual silylation on the oxygen terminus is not clear. The direct C-silylation of the lithium enolates of α,β-disubstituted esters is not possible, except with ethyl cyclopropanecarboxylate and ethyl cyclobutanecarboxylate.[4]

The α-(diphenylmethylsilyl) esters have been shown to be vinyl dication equivalents 3, and as such are precursors to terminal olefins and deuterated olefins,[5] 1,1-disubstituted olefins[6] and tri- and tetrasubstituted olefins.[7] They are precursors to β-ketosilanes and ketones,[8] wherein the overall transformation results in an ester to ketone conversion. They can also be deprotonated and the enolate anion condensed with aldehydes and ketones to give α,β-unsaturated esters,[9] in particular α-alkylated-α,β-unsaturated esters.[10] Their γ-lactone counterparts, α-(diphenylmethylsilyl)-γ-butyrolactone 4a and α-(diphenylmethylsilyl)-γ-valerolactone 4b, are precursors to 4-oxo acids,[11] 1,4-diketones[12] and α-ylidene-γ-lactones.[13]

$$\begin{array}{c} RCHCO_2Et \\ | \\ Ph_2SiMe \end{array} \quad \equiv \quad RCH{=}C^{+}_{+}$$

3

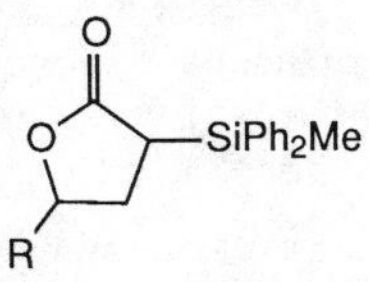

4a R = H
4b R = Me

1. Department of Chemistry, University of Puerto Rico, Rio Piedras, Puerto Rico 00931.
2. Watson, S. C; Eastham, J. F. *J. Organomet. Chem.* **1967**, *9*, 165.
3. Larson, G. L.; Fuentes, L. M. *J. Am. Chem. Soc.* **1981**, 103, 2418.
4. Larson, G. L.; Cruz de Maldonado, V. unpublished results.
5. Cruz de Maldonado, V.; Larson, G. L. *Synth. Commun.* **1983**, *13*, 1163.
6. Larson, G. L.; Hernandez, D. *Tetrahedron Lett.* **1982**, *23*, 1035.
7. Hernandez, D.; Larson, G. L. *J. Org. Chem.* **1984**, *49*, 4285.
8. (a) Larson, G. L.; Montes de Lopez-Cepero, I.; Torres, L. E. *Tetrahedron Lett.* **1984**, *25*, 1673; (b) Larson, G. L.; Hernandez, D.; Montes de Lopez-Cepero, I.; Torres, L. E. *J. Org. Chem.* **1985**, *50*, 5260.
9. Larson, G. L.; Quiroz, F.; Suarez, J. *Synth. Commun.* **1983**, *13*, 833.
10. Larson, G. L.; Fernandez de Kaifer, C.; Seda, R.; Torres, L. E.; Ramirez, J. R. *J. Org. Chem.* **1984**, *49*, 3385.
11. Fuentes, L. M.; Larson, G. L. *Tetrahedron Lett.* **1982**, *23*, 271.
12. Betancourt de Perez, R. M.; Fuentes, L. M.; Larson, G. L.; Barnes, C. L.; Heeg, M. J. *J. Org. Chem.* **1986**, *51*, 2039.
13. Larson, G. L.; Betancourt de Perez, R. *J. Org. Chem.* **1985**, *50*, 5257.

Appendix

Chemical Abstracts Nomenclature (Collective Index Number);

(Registry Number)

2-Methyl-2-undecene: 2-Undecene, 2-methyl- (9); (56888-88-1)

Ethyl 2-(diphenylmethylsilyl)decanoate: Decanoic acid, 2-(methyldiphenylsilyl)-, ethyl ester (11); (89638-16-4)

Ethyl decanoate: Decanoic acid, ethyl ester (8,9); (110-38-3)

Diphenylmethylchlorosilane: Silane, chloromethyldiphenyl- (8,9); (144-79-6)

N-BENZYL-N-METHOXYMETHYL-N-(TRIMETHYLSILYL)METHYLAMINE
AS AN AZOMETHINE YLIDE EQUIVALENT:
2,6-DIOXO-1-PHENYL-4-BENZYL-1,4-DIAZABICYCLO[3.3.0]OCTANE
(Pyrrolo[3,4-c]pyrrole-1,3(2H,3aH)-dione, tetrahydro-
2-phenyl-5-(phenylmethyl)-, cis-)

A. $PhCH_2NH_2$ + $(CH_3)_3SiCH_2Cl$ ⟶ $(CH_3)_3SiCH_2NHCH_2Ph$

B. $(CH_3)_3SiCH_2NHCH_2Ph$ $\xrightarrow[K_2CO_3]{\substack{CH_2O \\ CH_3OH}}$ $(CH_3)_3SiCH_2\underset{\underset{CH_2Ph}{|}}{N}CH_2OCH_3$

C. $(CH_3)_3SiCH_2\underset{\underset{CH_2Ph}{|}}{N}CH_2OCH_3$ + [N-phenylmaleimide] $\xrightarrow[\text{ultrasound}]{\substack{\text{LiF} \\ CH_3CN}}$ [bicyclic product]

Submitted by Albert Padwa and William Dent.[1]
Checked by Bruce Lefker and Albert I. Meyers.

1. Procedure

A. *N-Benzyl-N-(trimethylsilyl)methylamine.* An oven-dried, 100-mL, one-necked, round-bottomed flask equipped with a magnetic stirring bar and a reflux condenser is charged with 12.58 g (0.1 mol) of chloromethyltrimethylsilane (Note 1). Benzylamine (Note 2) (33.1 g, 0.3 mol) is added with stirring and the resulting solution is heated at 200°C for 2.5 hr. At the end of this time a 0.1 N sodium hydroxide solution is added in order to hydrolyze the white organic salt that had formed. The mixture is extracted with ether and the ether layer is dried over magnesium sulfate and concentrated under reduced pressure. The residue is distilled under reduced pressure through a 6-inch Vigreux column to give 11.6-15.3 g (58-72%) of N-benzyl-N-(trimethylsilyl)methylamine, bp 68-72°C (0.7-0.8 mm) (Note 3).

B. *N-Benzyl-N-methoxymethyl-N-(trimethylsilyl)methylamine.* A 25-mL, round-bottomed flask equipped with a stirring bar is charged with 6.0 g (74 mmol) of a 37% aqueous formaldehyde solution (Note 4). The solution is cooled to 0°C and 10.0 g (51.7 mmol) of N-benzyl-N-(trimethylsilyl)methylamine is added dropwise with stirring. After the solution is stirred for 10 min at 0°C, 6 mL (0.15 mol) of methanol (Note 5) is added in one portion. Potassium carbonate (4.0 g) is added to the mixture to absorb the aqueous phase. The mixture is stirred for 1 hr, the non-aqueous phase is decanted, an additional 2.0 g of potassium carbonate is added, and the mixture is stirred at 25°C for 12 hr. Ether is added to the mixture and the solution is dried over potassium carbonate, filtered, and concentrated under reduced pressure (Note 6). The residue is distilled at reduced pressure to give 6.8-8.6 g (54-69%) of N-benzyl-N-methoxymethyl-N-(trimethylsilyl)methylamine as a colorless liquid, bp 77-80°C (0.5 mm) (Note 7).

C. *2,6-Dioxo-1-phenyl-4-benzyl-1,4-diazabicyclo[3.3.0]octane.* An oven-dried, 250-mL, one-necked, round-bottomed flask equipped with a magnetic stirring bar is charged with 10.0 g (0.042 mol) of N-benzyl-N-methoxymethyl-N-(trimethylsilyl)methylamine and 100 mL of anhydrous acetonitrile (Note 8). N-Phenylmaleimide (Note 9) (7.3 g, 0.042 mol) is added followed by 1.7 g (0.063 mol) of lithium fluoride (Note 10). The reaction mixture is sonicated (Note 11) for 3 hr and poured into 100 mL of water. The mixture is extracted with three 100-mL portions of ether. The organic extracts are combined and washed with 100 mL of saturated sodium chloride solution, dried over magnesium sulfate, filtered, and concentrated under reduced pressure. The residue is chromatographed on a silica gel column (300 g) using a 35% ethyl acetate-hexane mixture (ca. 1500 mL) as the eluant to give 9.2-9.6 g (72-75%) of 2,6-dioxo-1-phenyl-4-benzyl-1,4-diazabicyclo[3.3.0]octane as a pale yellow solid, mp 97-98°C (Note 12).

2. Notes

1. Chloromethyltrimethylsilane is purchased from Petrarch Systems, Inc. and is used without purification.

2. Benzylamine, purchased from Aldrich Chemical Company, Inc., is distilled and stored over potassium hydroxide.

3. The submitters report bp 89-90°C (5 mm). The ^{1}H NMR spectrum (CDCl$_3$, 90 MHz) is as follows δ: 0.10 (s, 9 H), 2.00 (s, 2 H), 3.78 (s, 2 H) and 7.28 (s, 5 H).

4. Formaldehyde (37% solution in water) is purchased from Aldrich Chemical Company, Inc. Sufficient aqueous 10% sodium hydroxide solution (1-5 drops) is added until the pH reaches 7.

5. Purified grade methanol, purchased from Fisher Scientific Company, is used.

6. The submitters found it easier to pump down the crude mixture overnight under reduced pressure to ensure that all the methanol is removed. If not, the residue tends to bump uncontrollably upon distillation.

7. The spectral properties are as follows: IR (neat) cm^{-1}: 3095, 3064, 3030, 2900, 1605, 1495, 1450, 1422, 1385, 1362, 1245, 1070, 925, 845, 740 and 700; ^{1}H NMR (CDCl$_3$, 90 MHz) δ: 0.10 (s, 9 H), 2.13 (s, 2 H), 3.20 (s, 3 H), 3.72 (s, 2 H), 3.95 (s, 2 H) and 7.22 (m, 5 H).

8. Anhydrous acetonitrile, purchased from Aldrich Chemical Company, Inc., is distilled over calcium hydride and stored over 4 Å molecular sieves.

9. N-Phenylmaleimide is purchased from Aldrich Chemical Company, Inc., and used without purification.

10. Lithium fluoride is purchased from Fisher Scientific Company.

11. A Branson ultrasonic cleaner (2.8 liter, 13 x 23 x 10 cm), purchased from Fisher Scientific Company, is used for sonication. Without sonication, the yield drops by ca. 10-15%.

12. The spectral properties are as follows: IR (neat) cm^{-1}: 3145, 3000, 2950, 2900, 2800, 1760, 1700, 1575, 1490, 1445, 1380, 1310, 1200, 1155, 880, 840, 740 and 700; ^{1}H NMR (CDCl$_3$, 90 MHz) δ: 2.3-2.7 (m, 2 H), 3.2-3.6 (m, 4 H), 3.60 (s, 2 H) and 7.0-7.7 (m, 10 H).

3. Discussion

The preparation of pyrrolidines has received extensive attention by synthetic chemists in recent years, in part due to the interesting biological activities exhibited by several polysubstituted pyrrolidines.[2] Little attention has been given to one of the most conceptually simple ways of pyrrolidine formation: a 1,3-dipolar cycloaddition of an azomethine ylide with an olefin. This is not surprising since few methods exist for the preparation of nonstabilized azomethine ylides.[3-13] Silyl-substituted amines of Type 1 represent conjunctive reagents which can be considered as the equivalent of a nonstabilized azomethine ylide. These reagents have recently been found to undergo 1,3-dipolar cycloaddition to olefins to give pyrrolidine derivatives in good yield.[14-16] The present procedure provides a convenient route for the synthesis of a variety of five-membered ring nitrogen heterocycles using different dipolarophiles. Some representative examples are given in Table I. Advantages of the present method over existing methodologies include mild conditions, high yield and simplicity of the cycloaddition. Trimethylsilyl triflate or trimethylsilyl iodide can also be used.[12] However, these reagents are expensive and require longer reaction times.

We have found that sonication of the reaction mixture decreases the time needed for reaction and also substantially increases the yield. This is probably related to an increase in the solubility of lithium fluoride in acetonitrile or is a consequence of surface effects on the metal.

N-Benzyl-N-methoxymethyl-N-(trimethylsilyl)methylamine undergoes stereospecific cycloaddition with dimethyl maleate and fumarate. The cycloaddition behavior of an unsymmetrically substituted α-methoxysilylamine has also been examined and found to occur with high overall regioselectivity. The stereospecificity and regioselectivity of the reaction is consistent with a concerted 1,3-dipolar cycloaddition reaction.[17]

1. Department of Chemistry, Emory University, Atlanta, GA 30322.
2. "The Alkaloids, A Specialist Periodical Report", The Royal Society of Chemistry: London, 1983; Vol. 13.
3. Vedejs, E.; Martinez, G. R. *J. Am. Chem. Soc.* **1979**, *101*, 6452; **1980**, *102*, 7993.
4. Achiwa, K.; Sekiya, M. *Tetrahedron Lett.* **1982**, *23*, 2589; *Heterocycles*, **1983**, *20*, 167; *Chem. Lett.* **1981**, 1213.
5. Achiwa, K.; Motoyama, T.; Sekiya, M. *Chem. Pharm. Bull.* **1983**, *31*, 3939.
6. Terao, Y.; Imai, N.; Achiwa, K. Sekiya, M. *Chem. Pharm. Bull.* **1982**, *30*, 3167.
7. Tsuge, O.; Kanemasa, S.; Kuraoka, S.; Takenaka, S. *Chem. Lett.* **1984**, 279; **1984**, 281.
8. Tsuge, O.; Kanemasa, S.; Hatada, A.; Matsuda, K. *Chem. Lett.* **1984**, 801; Tsuge, O.; Kanemasa, S.; Takenaka, S. *Bull. Chem. Soc. Jpn.* **1983**, *56*, 2073.
9. Tsuge, O.; Oe, K.; Kawaguchi, N. *Chem. Lett.* **1981**, 1585.
10. Tsuge, O.; Ueno, K. *Heterocycles* **1983**, *20*, 2133; **1982**, *19*, 1411.
11. Grigg, R.; Basanagoudar, L. D.; Kennedy, D. A.; Malone, J. F.; Thianpatanagul, S. *Tetrahedron Lett.* **1982**, *23*, 2803; Grigg, R.; Gunaratne, H. Q. N.; Kemp, J. *Tetrahedron Lett.* **1984**, *25*, 99.
12. Hosomi, A.; Sakata, Y.; Sakurai, H. *Chem. Lett.* **1984**, 1117.
13. Chen, S. F.; Ullrich, J. W.; Mariano, P. S. *J. Am. Chem. Soc.* **1983**, *105*, 6160.
14. Padwa, A.; Dent, W. *J. Org. Chem.* **1987**, *52*, 235.
15. Padwa, A.; Dent, W.; Nimmesgern, H.; Venkatramanan, M. K.; Wong, G. S. K. *Chem. Ber.* **1986**, *119*, 813.

16. Parker, K. A.; Cohen, I. D.; Padwa, A.; Dent, W. *Tetrahedron Lett.* **1984**, *25*, 4917.

17. "1,3-Dipolar Cycloaddition Chemistry", Padwa, A., Ed.; Wiley: New York, 1984.

Appendix

Chemical Abstracts Nomenclature (Collective Index Number); (Registry Number)

2,6-Dioxo-1-phenyl-4-benzyl-1,4-diazabicyclo[3.3.0]octane:
Pyrrolo[3,4-c]pyrrole-1,3(2H,3aH)-dione, tetrahydro-2-phenyl-
5-(phenylmethyl)-, cis- (11): (87813-00-1)

N-Benzyl-N-methoxymethyl-N-(trimethylsilyl)methylamine: Benzenemethanamine,
N-(methoxymethyl)-N-[(trimethylsilyl)methyl]- (11); (93102-05-7)

N-Benzyl-N-(trimethylsilyl)methylamine: Benzenemethanamine,
N-[(trimethylsilyl)methyl]- (9); (53215-95-5)

N-Phenylmaleimide: Maleimide, N-phenyl- (8); 1 H-Pyrrole-2,5-dione,
1-phenyl- (9); (941-69-5)

TABLE

CYCLOADDITION OF **1** WITH ELECTRON-DEFICIENT DIPOLAROPHILES

Dipolarophile	Product	% Yield
(Z)-NC-CH=CH-CN	1-benzyl-3,4-dicyanopyrrolidine	90
(Z)-MeO₂C-CH=CH-CO₂Me (dimethyl maleate)	1-benzyl-3,4-bis(methoxycarbonyl)pyrrolidine (cis)	90
(E)-MeO₂C-CH=CH-CO₂Me (dimethyl fumarate)	1-benzyl-3,4-bis(methoxycarbonyl)pyrrolidine (trans)	90
$H_2C=CH-SO_2Ph$	1-benzyl-3-(phenylsulfonyl)pyrrolidine	92
PhCHO	3-benzyl-5-phenyl-1,3-oxazolidine	80
Ph₂C=S	3-benzyl-5,5-diphenyl-1,3-thiazolidine	91

ALKYLATIONS USING HEXACARBONYL(PROPARGYLIUM)DICOBALT SALTS:

2-(1-METHYL-2-PROPYNYL)CYCLOHEXANONE

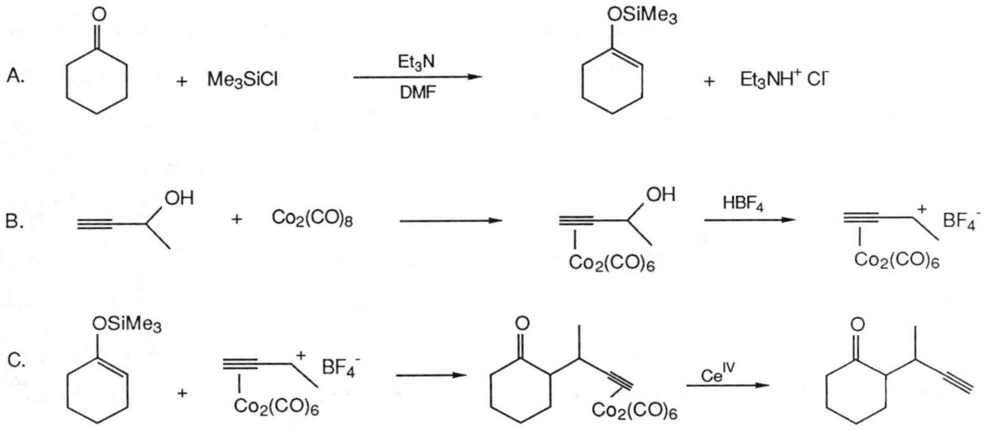

Submitted by Valsamma Varghese, Manasi Saha, and Kenneth M. Nicholas.[1]
Checked by T. V. RajanBabu, Leslie G. Upchurch, and Bruce E. Smart.

1. Procedure

Caution! Dicobalt octacarbonyl is highly toxic and air sensitive. All operations with this reagent should be carried out in an inert atmosphere and in a well-ventilated hood.

A. *1-Trimethylsiloxycyclohexene.*[2] A 1-L, three-necked, round-bottomed flask is equipped with a magnetic stirring bar, rubber septum, and reflux condenser fitted with a nitrogen gas inlet tube which is attached to a mineral oil bubbler. The system is flushed with nitrogen, flame dried, and while the system is maintained under a static pressure of nitrogen, the flask is charged with 300 mL of dry dimethylformamide (Note 1) and 110.3 g (1.1 mol) of triethylamine (Note 2); 58.3 g (0.54 mol) of chlorotrimethylsilane (Note 3) is added by syringe. Cyclohexanone (40.0 g, 0.41 mol) (Note 4) is added and the mixture is refluxed with stirring for 48 hr. After the flask is cooled to room temperature, the contents are poured into 600 mL of pentane. The resulting mixture is transferred to a separatory funnel and washed with three 500-mL portions of cold aqueous sodium bicarbonate. The organic layer is washed rapidly in succession with 200 mL of cold 1.5 N hydrochloric acid and 200 mL of cold aqueous sodium bicarbonate. The pentane solution is dried over sodium sulfate, filtered, and concentrated by rotary evaporation. The crude product is distilled through a short Vigreux column to give 53-54 g (76-77%) of 1-trimethylsiloxycyclohexene as a colorless liquid, bp 75-80°C (20-21 mm) (Note 5).

B. *Hexacarbonyl(1-methyl-2-propynylium)dicobalt tetrafluoroborate*, **1**. A 2-L, two-necked, round-bottomed flask fitted with a magnetic stirring bar, stopper, and gas inlet T-tube which is attached to a mineral oil bubbler is flame dried under a flow of nitrogen. The flask is charged with 200 mL of dry dichloromethane (Note 6) and 13.0 g (0.185 mol) of 3-butyn-2-ol (Note 7). After the mixture is stirred for 15 min, 65.0 g (0.19 mol) of dicobalt octacarbonyl (Note 8) is added in portions over a few minutes while maintaining a slow stream of nitrogen. Vigorous gas evolution (carbon monoxide!) is observed. The mixture is stirred for 4-5 hr, and the solvent is then removed under reduced pressure (20-25 mm). The residual solid (alkyne)-$Co_2(CO_6)$ complex is dissolved in 40 mL of propionic anhydride under nitrogen and cooled to -45°C in a dry ice/acetonitrile bath. Tetrafluoroboric acid-dimethyl etherate (37.3 g, 0.28 mol) (Note 9) is added with stirring. After

30 min, 600-800 mL of anhydrous diethyl ether is added with continuous stirring. The burgundy red salt which precipitates is isolated by filtration under a flow of nitrogen (Note 10) and is thoroughly washed with anhydrous diethyl ether to give 60-61 g (76-77%) of hexacarbonyl(1-methyl-2-propynylium)dicobalt tetrafluoroborate. This material is used immediately in the following step.

C.   2-(1-Methyl-2-propynyl)cyclohexanone. A 2-L, two-necked, round-bottomed flask is equipped with a magnetic stirring bar, stopper, and pressure-equalizing dropping funnel fitted with a gas inlet T-tube that is connected to a mineral oil bubbler. The flask is flushed with nitrogen and charged with 150 mL of dry dichloromethane (Note 6) and 60.0 g (0.141 mol) of the salt from Part B. The mixture is stirred and cooled to -78°C in a dry ice/2-propanol bath, and 23.9 g (0.141 mol) of 1-trimethylsiloxycyclohexene (Part A) is added dropwise over a few minutes. The mixture is stirred at -78°C for 4 hr. After the solution is warmed to room temperature, dichloromethane is removed under reduced pressure and replaced with 400 mL of acetone. The dark red solution of the alkyne complex is cooled to -78°C and 175 g (0.32 mol) of ceric ammonium nitrate (Note 11) is added in portions. The mixture is stirred until the gas evolution (carbon monoxide!) ceases (ca. 4 hr) (Note 12). The reaction mixture is warmed to room temperature, poured into 1 L of saturated brine solution, and extracted with four 250-mL portions of diethyl ether. The combined ether extracts are dried over magnesium sulfate, filtered, and concentrated on a rotary evaporator. The residual red oil is distilled at reduced pressure to afford 15.0-15.2 g (71-72%) of 2-(1-methyl-2-propynyl)cyclohexanone as a pale yellow liquid, bp 57-60°C (10 mm) (Note 13).

2. Notes

1. Dimethylformamide, obtained from Aldrich Chemical Company, Inc., was vacuum distilled from calcium hydride, bp 44°C (25 mm), and stored over 3 Å molecular sieves.

2. Triethylamine, obtained from the Aldrich Chemical Company, Inc., was distilled from potassium hydroxide prior to use.

3. Chlorotrimethylsilane, obtained from the Aldrich Chemical Company, Inc., was redistilled from calcium hydride before use.

4. Cyclohexanone was purchased from the Aldrich Chemical Company, Inc., redistilled, and stored over 4 Å molecular sieves.

5. The product is over 99.5% pure by GLPC (6' x 1/8" 3% SP 2100 on 100-120 mesh Supelcoport column) and has the following spectral characteristics: ^{1}H NMR (CDCl$_3$) δ: 0.21 (s, 9 H), 1.55 (m, 2 H), 1.69 (m, 2 H), 2.05 (br d, 4 H), 4.88 (br s, 1 H).

6. Dichloromethane, obtained from the Aldrich Chemical Company, Inc., was distilled from calcium hydride and stored over 4 Å molecular sieves.

7. 3-Butyn-2-ol was obtained from the Aldrich Chemical Company, Inc., and used without further purification.

8. Dicobalt octacarbonyl was obtained from Alfa Products, Morton/Thiokol, Inc. It is best weighed in a nitrogen-filled polyethylene glove bag or in a dry box.

9. Tetrafluoroboric acid-dimethyl etherate (d 1.38 g/mL) was purchased from the Aldrich Chemical Company, Inc. The submitters note that a tetrafluoroboric acid/acetic acid mixture, which is prepared by carefully adding 49% aqueous tetrafluoroboric acid (50 g, 0.28 mol) to ice-cold acetic anhydride (30.6 g, 0.30 mol), also can be used.

10. The filtration under nitrogen is conveniently carried out in a Schlenk filter flask.[3]

11. Ceric ammonium nitrate was obtained from the Aldrich Chemical Company, Inc.

12. The disappearance of the dark red (alkyne)$Co_2(CO)_6$ complex can be monitored by TLC on silica gel with a 1:9 diethyl ether:petroleum ether solvent mixture.

13. The product is obtained as a 2:1 diastereomeric mixture and is over 99% pure by GLPC (6' x 1/8" 3% SP 2100 on 100-120 mesh Supelcoport column). It has the following spectral characteristics: IR (CCl_4) 1710 cm^{-1}; ^{1}H NMR ($CDCl_3$) δ: 0.8-2.9 (br envelope, 10 H), 1.05 (d, 3 H, J = 7, minor diastereomer), 1.10 (d, 3 H, J = 7, major diastereomer), 2.15 (s, 1 H, both diastereomers); ^{13}C NMR ($CDCl_3$) δ: 16.3, 19.2, 24.2, 25.6, 24.7, 27.1, 28.4, 30.7, 41.7, 41.9, 54.1, 54.8, 68.2, 69.6, 86.3, 87.5, 209.7, 210.3; MS (70 eV) m/e 150, 121 (100%).

3. Discussion

In addition to their reactions with trimethylsilyl enol ethers, (propargylium)$Co_2(CO)_6^+$ complexes react with a variety of other mild carbon nucleophiles including activated aromatic compounds,[4] β-dicarbonyl compounds,[5] other enol derivatives (enol acetates and ketones directly),[6] allylsilanes,[7] and alkyl- and alkynyl-aluminum reagents.[8,9] These reactions provide a flexible means to introduce the synthetically versatile propargyl function. Key features of propargylations using these complexes are: 1) ready introduction and removal of the activating and directing -$Co_2(CO)_6$ group; 2) regiospecific attack by nucleophiles at the carbon α to the coordinated alkynyl group, giving propargyl products only (no allenic co-products); and 3) very mild reaction conditions and good overall yields.

The method reported here appears to be the one of choice for the dependable, efficient α-propargylation of ketones. It can be applied to propargylate ketones regioselectively at either the less substituted α-position (via the trimethylsilyl enol ether) or the more substituted α-position (using the enol acetate or even the ketone directly[6]). The resulting α-propargylated ketones are very useful synthetic intermediates. They have been converted to chromanols,[10] furans,[11] other heterocycles,[11] and cyclohexenones,[12] and they undergo regiospecific hydration to 1,4-diketones which, in turn, can be converted to cyclopentenones.[13-15] More classical indirect ketone propargylations generally give low yields with substantial co-production of allenic by-products, as with enamine[10,16] or acetoacetic ester propargylations.[11,17] Direct coupling of ketone enolates with propargyl halides or tosylates have rarely been attempted and can be expected to suffer the same limitations.

This preparation of 2-(1-methyl-2-propynyl)cyclohexanone appears to be the first reported.

1. Department of Chemistry, University of Oklahoma, Norman, OK 73019.
2. House, H. O.; Czuba, L. J.; Gall, M.; Olmstead, H. D. *J. Org. Chem.* **1969**, *34*, 2324.
3. Shriver, D. F. "The Manipulation of Air-Sensitive Compounds," McGraw Hill: New York 1969.
4. Lockwood, R. F.; Nicholas, K. M. *Tetrahedron Lett.* **1977**, 4163.
5. Hodes, H. D.; Nicholas, K. M. *Tetrahedron Lett.* **1978**, 4349.

6. Nicholas, K. M.; Mulvaney, M.; Bayer, M. *J. Am. Chem. Soc.* **1980**, *102*, 2508.
7. O'Boyle, J. E.; Nicholas, K. M. *Tetrahedron Lett.* **1980**, *21*, 1595.
8. Padmanabhan, S.; Nicholas, K. M. *J. Organometal Chem.* **1981**, *212*, 115.
9. Padmanabhan, S.; Nicholas, K. M. *Tetrahedron Lett.* **1983**, *24*, 2239.
10. Dufey, P. *Bull. Soc. Chim. Fr.* **1968**, 4653.
11. Schulte, K. E.; Reisch, J.; Bergenthal, D. *Chem. Ber.* **1968**, *101*, 1540.
12. Caine, D.; Tuller, F. N. *J. Org. Chem.* **1969**, *34*, 222.
13. Stork, G., Borch, R.; *J. Am. Chem. Soc.* **1964**, *86*, 935.
14. Padmanabhan, S., Nicholas, K. M. *Synth. Commun.* **1980**, *10*, 503.
15. Saha, M.; Nicholas, K. M. *Isr. J. Chem.* **1984**, *24*, 105.
16. Opitz, G. *Justus Liebigs Ann. Chem.* **1961**, *650*, 122.
17. Crombie, L.; Mackenzie, K. *J. Chem. Soc.* **1958**, 4417.

Appendix

Chemical Abstracts Nomenclature (Collective Index Number); (Registry Number)

1-Trimethylsiloxycyclohexene: Silane, (1-cyclohexen-1-yloxy)trimethyl- (8,9); (6651-36-1)

Hexacarbonyl(1-methyl-2-propynylium)dicobalt tetrafluoroborate: Cobalt(1+), hexacarbonyl[μ-[2,3,-η:2,3-η)-1-methyl-2-propynylium]]di-, (Co-Co), tetrafluoroborate(1-) (10); (62866-98-2)

3-Butyn-2-ol: 3-Butyn-2-ol, ($\pm$)- (10); (65337-13-5)

Dicobalt octacarbonyl: Cobalt, octacarbonyldi-, (Co-Co) (8,9); (15226-74-1)

Tetrafluoroboric acid-dimethyl etherate: Borate(1-), tetrafluoro-, hydrogen, compd. with oxybis[methane] (1:1) (10); (57969-83-9)

Ceric ammonium nitrate: Cerate(2-), hexanitrato-, diammonium (8); Cerate(2-), hexakis(nitrato-O)-, diammonium, (OC-6-11)- (9); (16774-21-3)

ETHYNYL p-TOLYL SULFONE

(Benzene, 1-(ethynylsulfonyl)-4-methyl-)

A. CH$_3$—C$_6$H$_4$—SO$_2$Cl + Me$_3$Si—C≡C—SiMe$_3$ $\xrightarrow{\text{AlCl}_3 / \text{CH}_2\text{Cl}_2}$

CH$_3$—C$_6$H$_4$—SO$_2$—C≡C—SiMe$_3$

B. CH$_3$—C$_6$H$_4$—SO$_2$—C≡C—SiMe$_3$ $\xrightarrow{\text{K}_2\text{CO}_3, \text{KHCO}_3 / \text{CH}_3\text{OH}, \text{H}_2\text{O}}$ CH$_3$—C$_6$H$_4$—SO$_2$—C≡C—H

Submitted by Liladhar Waykole and Leo A. Paquette.[1]
Checked by Dirk A. Heerding and Larry E. Overman.

1. Procedure

A. *p-Tolyl 2-(trimethylsilyl)ethynyl sulfone.* In a flame-dried, 500-mL, three-necked, round-bottomed flask fitted with a nitrogen inlet and glass stoppers are placed 200 mL of dry dichloromethane (Note 1) and 29.4 g (0.22 mol) of freshly powdered anhydrous aluminum chloride. After the addition of p-toluenesulfonyl chloride (41.9 g, 0.22 mol), the resulting dark brown mixture is shaken occasionally for 20 min at room temperature.

A 1-L, three-necked, round-bottomed flask equipped with a 500-mL addition funnel and a Teflon-coated stirring bar is flame-dried under a stream of dry nitrogen. The flask is charged with bis(trimethylsilyl)acetylene (34.0 g, 0.20 mol) (Note 2) and dry dichloromethane (200 mL) (Note 1) and the solution is cooled to 0°C in an ice-water bath.

The p-toluenesulfonyl chloride-aluminum chloride complex is quickly filtered through a glass-wool plug (Note 3) into the addition funnel. The residue is washed rapidly with an additional 50 mL of dry dichloromethane and the funnel is quickly stoppered. The complex is added dropwise during 1 hr to the cold (0°C), magnetically stirred silylacetylene solution. Upon completion of the addition, the reaction mixture is allowed to warm to room temperature and is stirred for an additional 12 hr. The mixture is hydrolyzed by pouring it into a slurry of 20% hydrochloric acid (200 mL) and ice (200 g) (Note 4). The organic layer is separated, washed twice with water (150 mL), and dried over anhydrous sodium sulfate. Removal of solvent in a rotary evaporator gives a brown solid (Note 5) which is recrystallized from light petroleum ether (bp 40-60°C) to yield 39.7-40.4 g (79-80%) of p-tolyl 2-(trimethylsilyl)ethynyl sulfone as white crystals, mp 81-82°C (Note 6).

B. *Ethynyl p-tolyl sulfone*. A 1-L, three-necked, round-bottomed flask equipped with a thermometer, 500-mL addition funnel, nitrogen inlet, and Teflon-coated magnetic stirring bar is charged with p-tolyl 2-(trimethylsilyl)ethynyl sulfone (25.2 g, 0.1 mol) and 300 mL of reagent grade methanol. After the mixture is stirred for 30 min, a clear solution is obtained. In the addition funnel is placed 350 mL of an aqueous solution containing potassium carbonate (6.2×10^{-3} M) and potassium bicarbonate (6.2×10^{-3} M); this buffer is added at a rate to maintain the reaction temperature at 30°C (Notes 7 and 8). The mixture is diluted with water (200 mL), and extracted with four 100-mL portions of chloroform. The combined organic phases are washed three times with water (100 mL) and twice with brine (100 mL) prior to drying over anhydrous sodium sulfate. Removal of solvent under reduced pressure leaves a creamy white solid, which is purified either by recrystallization from ethyl acetate-petroleum ether or by silica gel chromatography using 10% ethyl acetate in petroleum ether as eluant (Note 9). There is obtained 15.0 g (83%) of colorless crystals, mp 74-75°C (Notes 10 and 11).

2. Notes

1. The submitters used dichloromethane freshly distilled from powdered calcium hydride.

2. This reagent was obtained from Petrarch Systems, Inc., Bartram Road, Bristol, PA 19007.

3. This mixture is rather hygroscopic and must be maintained under a nitrogen atmosphere as much as possible.

4. Stirring facilitates the hydrolysis. The reaction mixture should be added relatively slowly since the decomposition is exothermic.

5. Material of this purity may be used directly in the ensuing step. However, lower yields are realized.

6. Earlier citations[2,3] report mp 81-82°C. This product has the following spectral properties: IR (KBr) cm^{-1}: 2124, 1338, 1164, 854, 779; ^1H NMR (CDCl$_3$) δ: 0.22 (s, 9 H), 2.48 (s, 3 H), 7.40 (d, 2 H, J = 9), 7.91 (d, 2 H, J = 9). MS (CI, 70 eV, isobutane) 253 (M + 1, 100). Anal. Calcd. for $C_{12}H_{16}O_2SSi$: C, 57.10; H, 6.40; S, 12.70. Found: C, 57.84; H, 5.88; S, 12.85. The checkers found that treatment of the crude solid with activated charcoal is required to obtain colorless product.

7. The checkers found that the reaction is complete immediately after addition if the temperature is maintained accurately at 30°C. The reaction rate is dramatically dependent on the reaction temperature. The submitters report that considerable resinous material is obtained if the temperature goes above 30°C.

8. This period of reaction may vary depending on the scale of the reaction. Progress may be easily followed by isolating aliquots and obtaining ^{1}H NMR spectra. The disappearance of the trimethylsilyl singlet is the observable diagnostic.

9. The checkers found that an impurity with a characteristic ^{1}H NMR singlet at 3.74 ppm is readily removed by recrystallization, but cannot be removed by chromatography. They also report that small amounts of this impurity are formed during flash chromatography on silica gel.

10. The ^{1}H NMR spectral characteristics of this sulfone are as follows (CDCl$_3$) δ: 2.47 (s, 3 H), 3.52 (s, 1 H), 7.38 (dd, 2 H, J = 8.5, 0.6), 7.88 (d, 2 H, J = 8.5). Its IR spectrum (KBr) consists of the following bands (cm^{-1}): 3235, 2013, 1337, 1156. MS (CI, 70 eV, isobutane) 181 (M + 1, 100). Anal. Calcd. for C$_9$H$_8$SO$_2$: C, 59.97; H. 4.47; S, 17.79. Found: C, 59.20; H, 4.55; S, 17.52.

11. Further purification can be achieved if desired by recrystallization of this material from hexane-ethyl acetate (95:5). Shiny needles which melt at 75°C are thereby obtained.

3. Discussion

Interest in arylsulfonyl acetylenes arose initially because of their powerful Michael acceptor properties. Examples of facile nucleophilic addition involving thiolates (eq. 1),[4-7] amines,[2] cuprates (eq. 2),[8] malonate

(1)

$$PhSO_2C\equiv CH \xrightarrow[\text{2. PhSH}]{\text{1. } R_2CuLi} PhSO_2CH=CHR \xrightarrow[\text{2. } H^+]{\text{1. } R'_2CuLi} PhSO_2CH_2CH{\overset{R}{\underset{R'}{\diagup}}} \quad (2)$$

anions,[9] alkoxides,[10,11] hydroxylamines,[12,13] and azlactone enolates[14] abound. More recently, the dienophilic properties of this class of compounds have been used to advantage, e.g., use of the title compound as an acetylene synthon in Diels-Alder cycloadditions,[15,16] its [4+2] capture by N-methoxy-carbonylpyrrole in a first step toward the elusive 7-azanorbornadiene,[17] and its pivotal role in a synthesis of [4]-peristylane (eq. 3).[18] Ethynyl p-tolyl sulfone undergoes $EtAlCl_2$-catalyzed ene reactions with alkenes to give

1,4-dienyl p-tolyl sulfones (eq. 4).[19] Condensations with ynamines to give 2-amino-5-arylsulfinylfurans (eq. 5) have been reported.[20] α,β-Acetylenic sulfones also react with organolithium and Grignard reagents to give the correspondingly higher acetylene (eq. 6).[21]

$$\text{ArSO}_2\text{C}{\equiv}\text{CR} \xrightarrow{\text{R'Li or}}_{\text{R'MgBr}} \text{R'}-\text{C}{\equiv}\text{C}-\text{R} \qquad (6)$$

The procedures used most often for preparation of arylsulfonyl acetylenes involve oxidation of the corresponding ethynyl thio ether. The thio ethers are usually obtained via a two-step sequence beginning with two-fold thiophenoxide displacement of chloride ion from cis-1,2-dichloroethylene, followed by elimination with n-butyllithium in the resultant cis-1,2-bisarylthioethylene.[19] Less well known methods involve diazotization of 4-arylsulfonyl-5-aminoisoxazoles,[22] dehydrobromination of cis- and trans-2-bromovinyl phenyl sulfone with fluoride ion,[23] and oxidative elimination of β-(phenylseleno)vinyl sulfones.[24] The method described here, which bypasses the need for strongly basic conditions, is adapted from the work of Bhattacharya, Josiah, and Walton.[3] The simplicity and mildness of the method suggest that it may be broadly useful.

1. Department of Chemistry, The Ohio State University, Columbus, OH 43210.
2. Maioli, L.; Modena, G. *Ricerca. sci.* **1959**, *29*, 1931.
3. Bhattacharya, S. N.; Josiah, B. M.; Walton, D. R. M. *Organometal. Chem. Syn.* **1971**, *1*, 145.
4. Stirling, C. J. M. *J. Chem. Soc.* **1964**, 5856.
5. Truce, W. E.; Tichenor, G. J. W. *J. Org. Chem.* **1972**, *37*, 2391.
6. De Lucchi, O.; Marchioro, C.; Valle, G.; Modena, G. *J. Chem. Soc., Chem. Commun.* **1985**, 878.
7. De Lucchi, O.; Lucchini, V.; Marchioro, C.; Valle, G.; Modena, G. *J. Org. Chem.* **1986**, *51*, 1457.

8. Fiandanese, V.; Marchese, G.; Naeso, F. *Tetrahedron Lett.*, **1978**, 5131.
9. Eisch, J. J.; Behrooz, M.; Dua, S. K. *J. Organometal. Chem.* **1985**, *285*, 121.
10. Di Nunno, L.; Modena, G.; Scorrano, G. *J. Chem. Soc B.* **1966**, 1186.
11. van der Sluijs, M. J.; Stirling, C. J. M. *J. Chem. Soc., Perkin Trans. 2* **1974**, 1268.
12. Sanders, J. A.; Hovius, K.; Engberts, J. B. F. N. *J. Org. Chem.* **1974**, *39*, 2641.
13. Aurich, H. G.; Hahn, K. *Chem. Ber.* **1979**, *112*, 2769.
14. Steglich, W.; Wegmann, H. *Synthesis* **1980**, 481.
15. Davis, A. P.; Whitham, G. H. *J. Chem. Soc., Chem. Commun.* **1980**, 639.
16. Paquette, L. A.; Carr, R. V. C.; Bohm, M.; Gleiter, R. *J. Am. Chem. Soc.* **1980**, *102*, 1186.
17. Altenbach, H.-J.; Blech, B.; Marco, J. A.; Vogel, E. *Angew. Chem.* **1982**, *94*, 789; *Angew. Chem. Intern. Ed. Engl.* **1982**, *21*, 772.
18. Paquette, L. A.; Browne, A. R.; Doecke, C. W.; Williams, R. V. *J. Am. Chem. Soc.* **1983**, *105*, 4113; Paquette, L. A.; Fischer, J. W.; Browne, A. R.; Doecke, C. W. *J. Am. Chem. Soc.* **1985**, *107*, 686.
19. Snider, B. B.; Kirk, T. C.; Roush, D. M.; Gonzalez, D. *J. Org. Chem.* **1980**, *45*, 5015.
20. Himbert, G.; Kosack, S.; Maas, G. *Angew. Chem.* **1984**, *96*, 308; *Angew. Chem., Intern. Ed. Engl.* **1984**, *23*, 321.
21. Smorada, R. L.; Truce, W. E. *J. Org. Chem.* **1979**, *44*, 3444.
22. Beccalli, E. M.; Manfredi, A.; Marchesini, A. *J. Org. Chem.* **1985**, *50*, 2372.
23. Naso, F.; Ronzini, L. *J. Chem. Soc., Perkin Trans. 1* **1974**, 340.
24. Back, T. G.; Collins, S.; Kerr. R. G. *J. Org. Chem.* **1983**, *48*, 3077.

Appendix

Chemical Abstracts Nomenclature (Collective Index Number);
(Registry Number)

Ethynyl p-tolyl sulfone: Sulfone, ethynyl p-tolyl (8); Benzene, 1-(ethynylsulfonyl)-4-methyl- (9); (13894-21-8)

p-Tolyl 2-(trimethylsilyl)ethynyl sulfone: Silane, trimethyl [[(4-methylphenyl)sulfonyl]ethynyl]- (9); (34452-56-7)

Bis(trimethylsilyl)acetylene: Silane, 1,2-ethynediylbis[trimethyl- (9); (14630-40-1)

DIENOPHILE ACTIVATION VIA SELENOSULFONATION:
1-(BENZENESULFONYL)CYCLOPENTENE
(Benzene, (1-cyclopenten-1-ylsulfonyl)-)

A. $PhSO_2NHNH_2$ + $PhSeO_2H$ $\xrightarrow[25°C]{CH_2Cl_2}$ $PhSeSO_2Ph$ + N_2 + $2H_2O$

B. [cyclopentene] + $PhSeSO_2Ph$ $\xrightarrow{\text{1. hv, CCl}_4}_{\text{2. H}_2\text{O}_2,\ \text{CH}_2\text{Cl}_2}$ [1-(benzenesulfonyl)cyclopentene, SO_2Ph]

Submitted by Ho-Shen Lin, Michael J. Coghlan, and Leo A. Paquette.[1]
Checked by Tony Haight and Edwin Vedejs.

1. Procedure

A. Phenyl benzeneselenosulfonate. A 1-L, three-necked, round-bottomed flask equipped with a Teflon-coated magnetic stirring bar and a 250-mL addition funnel containing 17.2 g (100 mmol) of benzenesulfonyl hydrazide (Note 1) and 125 mL of dichloromethane is charged with 18.9 g (100 mmol) of phenylseleninic acid (Note 1) and 125 mL of dichloromethane. The seleninic acid is stirred at 25°C as the hydrazide slurry is added over 1 hr (Note 2). After an additional hour at 25°C, the reaction mixture is dried over anhydrous magnesium sulfate, filtered, and concentrated under reduced pressure. The residue is dissolved in 250 mL of hot methanol and the solution of selenosulfonate is cooled overnight at ~5°C in a refrigerator to induce crystallization. The yellow product which precipitates is filtered and recrystallized from methanol to afford 24.3-25.2 g (83-85%) of phenyl benzeneselenosulfonate, mp 56°C (Note 3).

B. *1-(Benzenesulfonyl)cyclopentene*. A one-necked, flat-bottomed, cylindrical flask (5 cm in diameter and 23 cm in height) equipped with a Teflon-coated magnetic stirring bar is charged in turn with phenyl benzeneselenosulfonate (15.0 g, 50.5 mmol), carbon tetrachloride (160 mL), and cyclopentene (11.1 mL, 126 mmol) (Note 4). The flask is equipped with a Friedrichs condenser and the stirred reaction mixure is blanketed with nitrogen. Following irradiation with a 150W sunlamp at room temperature for 45 min, the solution is transferred to a 500-mL, one-necked, round-bottomed flask and concentrated on a rotary evaporator.

A Teflon-coated magnetic stirring bar is placed atop the residue, which is dissolved in 140 mL of dichloromethane. The stirred solution is cooled in an ice-water bath to 0°C as 60 mL of 15% hydrogen peroxide is added dropwise via an addition funnel over 30 min (Note 5). Vigorous stirring is maintained at this temperture for 1.5 hr. The mixture is transferred to a 1-L separatory funnel, diluted with 400 mL of ethyl acetate and washed twice with 150-mL portions of water. The organic layer is dried over anhydrous magnesium sulfate, filtered, and freed of solvent under reduced pressure. The residual yellowish solid is dissolved in a small amount of dichloromethane and eluted with 5% ethyl acetate in dichloromethane through a column of 80 g of neutral alumina (activity III) to afford 8.28-9.19 g (79-87%) of colorless crystals, mp 65-66°C. ^{1}H NMR analysis shows this material to be of very high purity (Note 6).

2. Notes

1. Benzenesulfonyl hydrazide is available from the Fluka Chemical Company, 255 Oser Avenue, Hauppauge, NY 11788.

2. This addition time ensures a slow, steady evolution of nitrogen during admixture of both slurried reactants.

3. Although this selenosulfonate is temperature and light sensitive, it can be stored indefinitely at refrigerator temperatures in an opaque glass container.[2]

4. Cyclopentene was purchased from the Aldrich Chemical Company, Inc. and used without further purification.

5. The peroxide is added at such a rate that the mildly exothermic oxidation/elimination reaction is well controlled. Faster additon of hydrogen peroxide can result in uncontrollable foaming.

6. The product has the following spectral properties: IR (KBr) cm^{-1}: 3060, 2960, 2920, 2840, 1610, 1580, 1440, 1300, 1150, 1085, 935, 825, 745, 710, 680, 600; ^{1}H NMR (CDCl$_3$) δ: 1.8-2.2 (m, 2 H, CH$_2$C$\underline{H}_2$-CH$_2$), 2.2-2.6 (m, 4 H, C$\underline{H}_2$C=), 6.6 (br s, 1 H, =C$\underline{H}$), 7.3-8.0 (m, 5 H); m/z Calcd for C$_{11}$H$_{12}$O$_2$S: 208.0558; Found 208.0553. Anal. Calcd for C$_{11}$H$_{12}$O$_2$S: C, 63.43, H, 5.81. Found: C, 63.49; H, 5.83.

3. Discussion

Recent investigations into the chemistry of vinyl sulfones have revealed that they are versatile synthetic intermediates, serving either as dienophiles[3] or Michael acceptors.[4] Methods for the preparation of vinyl sulfones from unactivated olefins have customarily involved the catalyzed (boron trifluoride or benzoyl peroxide) addition of PhSO$_2$X (X = Cl, Br, I, or SePh), followed by elimination of HX.[5] However, when phenylsulfonyl halides are employed, yields are variable, reactions are frequently incomplete, and the Lewis acid or free-radical catalyst employed can potentially interfere with other functionality present. On the other hand, the selenosulfonation method, particularly when photochemically induced,[3,6] proceeds smoothly to completion in high yield and is compatible with several functional groups (Table I).[3] A further consequence of the trans disposition of the benzeneselenenyl and benzenesulfonyl groups is invariant elimination to give the α,β-unsaturated sulfone.

1. Department of Chemistry, The Ohio State University, Columbus, OH 43210.
2. Back, T. G.; Collins, S. *J. Org. Chem.* **1981**, *46*, 3249.
3. Kinney, W. A.; Crouse, G. D.; Paquette, L. A. *J. Org. Chem.* **1983**, *48*, 4986. Review: De Lucchi, O.; Modena, G. *Tetrahedron* **1984**, *40*, 2585.
4. Hamann, P. R.; Fuchs, P. L. *J. Org. Chem.* **1983**, *48*, 914; Pyne, S. G.; Spellmeyer, D. C.; Chen, S.; Fuchs, P. L. *J. Am. Chem. Soc.* **1982**, *104*, 5728; Cory, R. M.; Renneboog, R. M. *J. Org. Chem.* **1984**, *49*, 3898; Auvray, P.; Knochel, P.; Normant, J. F. *Tetrahedron Lett.* **1985**, *26*, 2329; Eisch, J. J.; Galle, J. E. *J. Org. Chem.* **1979**, *44*, 3277; De Lucchi, O.; Pasquato, L.; Modena, G. *Tetrahedron Lett.* **1984**, *25*, 3647; Ueno, Y.; Khare, R. K.; Okawara, M. *J. Chem. Soc., Perkin Trans. 1* **1983**, 2637; Ochiai, M.; Ukita, T.; Fujita, E. *J. Chem. Soc., Chem. Commun.* **1983**, 619; Donaldson, R. E.; Fuchs, P. L. *J. Am. Chem. Soc.* **1981**, *103*, 2108.
5. Skell, P. S.; McNamara, J. H. *J. Am. Chem. Soc.* **1957**, *79*, 85; Cristol, S. J.; Davies, D. I. *J. Org. Chem.* **1964**, *29*, 1282; Harwood, L. M.; Julia, M.; Le Thuillier, G. *Tetrahedron* **1980**, *36*, 2483; Böll, W. *Liebigs Ann. Chem.* **1979**, 1665. See also Craig, D.; Ley, S. V.; Simpkins, N. S.; Whitham, G. H.; Prior, M. J. *J. Chem. Soc., Perkin Trans. 1* **1985**, 1949.
6. Gancarz, R. A.; Kice, J. L. *J. Org. Chem.* **1981**, *46*, 4899.

TABLE

PHOTOINDUCED SELENOSULFONATION-ELIMINATION OF OLEFINS

Olefin	Product	Yield (%)
CH₂=CH-CH₂-CH₂-CH₂-CH₃	PhSO₂-CH=CH-CH₂-CH₂-CH₂-CH₃	62
CH₂=CH-CH₂-OPh	PhSO₂-CH=CH-CH₂-OPh	89
CH₂=CH-SiMe₃	PhSO₂-CH=CH-SiMe₃	84
CH₂=CH-CH(OCH₃)₂	PhSO₂-CH=CH-CH(OCH₃)₂	73
CH₂=CH-CH₂-CH₂-SiMe₃	PhSO₂-CH=CH-CH₂-CH₂-SiMe₃	93
CH₂=CH-CH₂-CH₂-CH₂-O-Si(tBu)Me₂	PhSO₂-CH=CH-CH₂-CH₂-CH₂-O-Si(tBu)Me₂	89
CH₂=CH-CH₂-O-Si(tBu)Me₂	PhSO₂-CH=CH-CH₂-O-Si(tBu)Me₂	85
bicyclic lactone with vinyl	bicyclic lactone with CH=CH-SO₂Ph	72
MeO₂C-cyclopentene (H)	MeO₂C-cyclopentene-SO₂Ph (H)	90
2,5-dihydrofuran	3-SO₂Ph-2,5-dihydrofuran	75

Appendix

Chemical Abstracts Nomenclature (Collective Index Number);

(Registry Number)

1-(Benzenesulfonyl)cyclopentene: Benzene, (1-cyclopenten-1-ylsulfonyl)- (10); (64740-90-5)

Phenyl benzeneselenosulfonate: Benzenesulfonoselenoic acid, Se-phenyl ester (9); (60805-71-2)

Benzenesulfonyl hydrazide: Benzenesulfonic acid, hydrazide (8,9); (80-17-1)

Phenylseleninic acid: Benzeneseleninic acid (8,9); (6996-92-5)

Cyclopentene (8,9); (142-29-0)

REDUCTIVE ANNULATION OF VINYL SULFONES: BICYCLO[4.3.0]NON-1-EN-4-ONE
(5H-Inden-5-one, 1,2,3,3a,4,6-hexahydro-)

A.

B.

Submitted by Ho-Shen Lin and Leo A. Paquette.[1]
Checked by Edward J. Adams and Edwin Vedejs.

1. Procedure

A. *4-Oxo-1-(benzenesulfonyl)-cis-bicyclo[4.3.0]non-2-ene.* A 250-mL, one-necked flask equipped with a Teflon-coated magnetic stirring bar, condenser, and nitrogen inlet tube is charged with 8.76 g (42.1 mmol) of 1-(benzenesulfonyl)cyclopentene (Note 1), 8.75 g (50.9 mmol) of 1-methoxy-3-(trimethylsiloxy)-1,3-butadiene and 8 mL of xylene (Note 2). The stirred reaction mixture is blanketed with nitrogen and heated in an oil bath at 123-125°C in the dark for 3 days. After the solution is cooled, it is diluted with 80 mL of tetrahydrofuran and 30 mL of 2 N hydrochloric acid and heated at reflux temperature for 24 hr. Most of the tetrahydrofuran is removed on a rotary evaporator. The residue is transferred to a 1-L separatory funnel and diluted with ether (200 mL) and dichloromethane (100 mL). The organic phase is washed with two 50-mL volumes of half-saturated sodium bicarbonate solution and a mixed solution of saturated sodium bicarbonate (10 mL) and half-saturated sodium chloride (40 mL). The organic layer is dried over anhydrous magnesium sulfate, filtered, and concentrated. The residue is subjected to flash column chromatography on silica gel (250 g). Elution with a mixture of ethyl acetate-dichloromethane-petroleum ether (1:25:25) returns 3.56 g (41%) of unreacted 1-(benzenesulfonyl)cyclopentene. Subsequent increase in the solvent polarity to 3:25:25 provides the cycloadduct as a yellowish solid. This material is dissolved in the minimum amount of dichloromethane to which is added 25 mL of ether; 4.33 g of colorless crystals precipitate. Concentration of the filtrate and crystallization from ether-petroleum ether afford an additional 0.77-1.34 g of light yellow crystals (combined yield of 44-49%) (Note 3).

B. *Bicyclo[4.3.0]non-1-en-4-one*. The preceding enone (5.64 g, 20.4 mmol) is dissolved with magnetic stirring in 120 mL of glacial acetic acid contained in a 500-mL, one-necked flask. Zinc powder (13.3 g, 0.203 mol) (Note 4) is introduced and the capped reaction mixture is stirred vigorously at room temperature for 1 hr. The zinc is removed by suction filtration through a Celite pad (Büchner funnel) and washed with 200 mL of ether. The combined filtrates are transferred to a 2-L separatory funnel, diluted with 300 mL of petroleum ether, and washed with 200 mL of water. The aqueous phase is reextracted with 300 mL of a 1:1 mixture of ether and petroleum ether. Finally the combined organic layers are washed with 200 mL of water and 200 mL of saturated sodium bicarbonate solution prior to drying over anhydrous magnesium sulfate and filtration. The solvents are removed on a rotary evaporator to leave a pale yellow oil which is purified by chromatography on silica gel (elution with 14% ethyl acetate in petroleum ether). There is isolated 1.81-1.98 g (65-71%) of the β,γ-enone as a colorless oil (Notes 5 and 6).

2. Notes

1. See the preceding procedure for preparation of this intermediate.

2. The reagents were purchased from the Aldrich Chemical Company, Inc.; the diene was used without further purification, and xylene was dried by azeotropic removal of water and distillation from calcium hydride. 1-Methoxy-3-trimethylsiloxy-1,3-butadiene can be prepared by the method of *Org. Synth.* **1983**, *61*, 147.

3. The product can be further purified by crystallization from dichloromethane and ether. The crystalline modification that is obtained melts at 122.5-126°C. Melting and resolidification provides a second modification that melts at 122.5-123.2°C. The product has the following spectral properties: IR (CH_2Cl_2) cm^{-1}: 1680, 1310, 1150, 1090; ^{1}H NMR ($CDCl_3$) δ: 1.27-1.49 (m, 1 H), 1.55-1.87 (m, 3 H), 1.95-2.15 (m, 1 H), 2.16-2.31 (m, 2 H), 2.59-2.76 (m, 1 H), 2.93-3.13 (m, 1 H), 6.08 (d, 1 H, J = 10.2), 6.49 (dd, 1 H, J = 10.0, 1.7), 7.53 (br t, 2 H, J = 7.3), 7.64 (br t, 1 H, J = 7.2), 7.83 (br d, 2 H, J = 7.2); ^{13}C NMR ($CDCl_3$) δ: 22.76, 32.34, 35.68, 37.91, 38.73, 70.46, 129.05, 129.64, 131.89, 134.19, 136.38, 143.54, 196.27; m/z Calcd for M^+-$C_6H_5SO_2$: 135.0801; Found 135.0835. Anal. Calcd for $C_{15}H_{16}O_3S$: C, 65.19; H, 5.84. Found: C, 65.28, H. 5.85.

4. Fresh Mallinckrodt zinc dust was used without purification.

5. The product has the following spectral properties: IR (CH_2Cl_2) cm^{-1}: 2960, 2880, 1710; ^{1}H NMR ($CDCl_3$) δ: 1.08-1.35 (m, 1 H), 1.43-1.68 (m, 1 H), 1.68-1.91 (m, 1 H), 1.91-2.15 (m, 2 H), 2.15-2.38 (m, 2 H), 2.38-2.65 (m, 2 H), 2.65-2.94 (m, 2 H), 5.38-5.53 (m, 1 H); ^{13}C NMR ($CDCl_3$) δ: 24.86, 29.78, 33.80, 39.09, 40.40, 45.07, 113.22, 146.32, 211.05; m/z Calcd for $C_9H_{12}O$: 136.0888; Found 136.0896.

6. The product may be isolated by distillation, although two complications arise. First, because of the volatility of the enone, some material loss is incurred (yields of 57-60% result), bp 68-78°C (19 mm). More critically, heating induces some equilibration (generally about 10-15%) to the α,β-enone isomer. Thus, distillation should be avoided if pure β,γ-enone is desired. The spectral properties of the conjugated ketone, which can be obtained in a pure state by silica gel chromatography, are: IR (CH_2Cl_2) cm^{-1}: 2950, 2875, 1670; ^{1}H NMR ($CDCl_3$) δ: 1.31-1.49 (m, 1 H), 1.57-2.08 (m, 5 H), 2.40 (dd, 1 H, J = 17.8, 7.4), 2.47-2.62 (m, 2 H), 2.72-2.84 (m, 1 H), 5.92 (dd, 1 H, J = 10.2, 2.1), 6.70 (dd, 1 H, J = 10.2, 3.5); m/z Calcd for $C_9H_{12}O$: 136.0888; Found 136.0864.

3. Discussion

As a group, annulation reactions have been exceedingly valuable to the synthetic organic chemist. Unfortunately, processes of this type involving simple alkenes and cycloalkenes are few. However, the facility with which unactivated olefins can be transformed into vinyl sulfones,[2,3] the high degree to which α,β-unsaturated sulfones are captured regioselectively by unsymmetrical dienes[4] such as those developed by Danishefsky,[5] and the ease with which reductive desulfonylation can be effected,[6,7] combine to permit convenient synthetic entry to substituted cyclohexenones. Several representative examples can be found in Table I.

Other variants on this theme are possible. Thus, if the initially-formed Diels-Alder adduct is directly ketalized as in 2, the derived α-sulfonyl carbanion can be alkylated. Reductive desulfonylation and acidic hydrolysis (with pyridinium p-toluenesulfonate, PPTS) then deliver a 4-substituted cyclohexenone (e.g., 3), which in many cases can be made to undergo further useful synthetic transformations (e.g., 4).[4]

The expandibility of the scheme allows one to prepare 4-substituted and 4,5-disubstituted 2(and 3)-cyclohexenones where the nature of the side chains can be widely varied.

1. Department of Chemistry, The Ohio State University, Columbus, Ohio 43210.
2. Carr, R. V. C.; Williams, R. V.; Paquette, L. A. J. Org. Chem. **1983**, 48, 4976 and pertinent references cited therein.
3. Lin, H.-S.; Coghlan, M. J.; Paquette, L. A. Org. Synth. **1988**, 67, 158.
4. Kinney, W. A.; Crouse, G. D.; Paquette, L. A. J. Org. Chem. **1983**, 48, 4986 and relevant references cited therein.

5. (a) Danishefsky, S.; Kitahara, T. *J. Am. Chem. Soc.* **1974**, *96*, 7807; (b) Danishefsky, S.; Yan, C. F.; McCurry, P. M., Jr. *J. Org. Chem.* **1977**, *42*, 1819; (c) Danishefsky, S.; Kitahara, T.; Yan, C. F.; Morris, J. *J. Am. Chem. Soc.* **1979**, *101*, 6996; (d) Danishefsky, S.; Yan, C. F.; Singh, R. K.; Gammill, R. G.; McCurry, P. M., Jr.; Fritsch, N.; Clardy, J. *J. Am. Chem. Soc.* **1979**, *101*, 7001.

6. Lansbury, P. T.; Erwin, R. W.; Jeffrey, D. A. *J. Am. Chem. Soc.* **1980**, *102*, 1602.

7. Trost, B. M.; Arndt, H. C.; Strege, P. E.; Verhoeven, T. R. *Tetrahedron Lett.* **1976**, 3477.

Appendix
Chemical Abstracts Nomenclature (Collective Index Number); (Registry Number)

1-(Benzenesulfonyl)cyclopentene: Benzene, (1-cyclopenten-1-ylsulfonyl)- (10); (64740-90-5)

1-Methoxy-3-(trimethylsiloxy)-1,3-butadiene: Silane, [(3-methoxy-1-methylene-2-propenyl)oxy]trimethyl- (9); (59414-23-2)

TABLE

REDUCTIVE ANNULATION OF VINYL SULFONES[4]

Starting material	Diene	Product	α,β:β,γ Ratio
PhSO$_2$–CH=CH–CH$_2$CH$_2$CH$_2$CH$_3$	1b	5-butyl-2-methylcyclohex-2-enone	100:0
PhSO$_2$–CH=CH–CH$_2$–OPh	1b	5-(phenoxymethyl)-2-methylcyclohex-2-enone	100:0
PhSO$_2$–CH=CH–CH(OCH$_3$)$_2$	1b	5-(dimethoxymethyl)-2-methylcyclohex-2-enone	60:40
cis-bicyclic lactone with CH=CH–SO$_2$Ph substituent	1a	bicyclic lactone bearing cyclohex-2-enone	0:100
3-(PhSO$_2$)-2,5-dihydrofuran	1b	fused bicyclic enone with CH$_3$	0:100
PhSO$_2$–CH=CH–CH$_2$CH$_2$–SiMe$_3$	1a	cyclohexenone with Me$_3$Si–CH$_2$CH$_2$– and –CH$_2$CH$_2$CH(CH$_3$)–(p-tolyl) substituents [a]	0:100

[a] Including an intermediate alkylation step.

ETHYL α-(HEXAHYDROAZEPINYLIDENE-2)ACETATE FROM O-METHYLCAPROLACTIM AND MELDRUM'S ACID

(Acetic acid, (hexahydro-2H-azepin-2-ylidene)-, ethyl ester, (Z)-)

Submitted by J. P. Celerier, E. Deloisy-Marchalant, G. Lhommet, and P. Maitte.[1]
Checked by Ting-Zhong Wang and Leo A. Paquette.

1. Procedure

A. *Isopropylidene α-(hexahydroazepinylidene-2)malonate.* In a 1-L, round-bottomed flask fitted with an efficient reflux condenser and equipped with a magnetic stirrer are placed 50.8 g (0.40 mol) of O-methylcaprolactim (Note 1), 57.6 g (0.40 mol) of Meldrum's acid (Note 2) and 0.25 g of nickel acetylacetonate monohydrate (Note 3) in 500 mL of anhydrous chloroform. The reaction mixture is refluxed for 12 hr. The solvent is removed with a rotary evaporator and the bright yellow precipitate is recrystallized from absolute ethanol to give 77-78 g (81-82%) of pale yellow crystals, mp 147-149°C (Note 4)

B. *Ethyl α-(hexahydroazepinylidene-2)acetate.* A solution of sodium ethoxide is prepared from 8.3 g (0.36 mol) of freshly cut sodium and 600 mL of freshly distilled absolute ethanol (Note 5) in a 1-L, round-bottomed flask equipped with a magnetic stirrer and fitted with a reflux condenser. To the stirred solution is added in one portion 71.7 g (0.30 mol) of freshly recrystallized isopropylidene α-(hexahydroazepinylidene-2)malonate. The mixture is refluxed and a white precipitate begins to appear. Refluxing is continued for 12 hr. The solvent is removed with a rotary evaporator and the white precipitate is placed in a 2-L beaker. Water (300 mL) is added and a 1 N hydrochloric solution is added dropwise to pH 6. The reaction mixture is extracted with four 100-mL portions of chloroform. The extracts are dried over anhydrous sodium sulfate and the solvent is removed with a rotary evaporator. The yellow solid residue is recrystallized from methanol to give 43-44 g (78-80%) of white powder, mp 55-56°C (Note 6).

2. Notes

1. O-Methylcaprolactim, (1-aza-2-methoxy-1-cycloheptene), is available from the Janssen Chimica Society (France) and from the Aldrich Chemical Company, Inc. It may be also prepared from ε-caprolactam and dimethyl sulfate.[2]

2. Meldrum's acid, (2,2-dimethyl-1,3-dioxane-4,6-dione), is available from the Janssen Chimica Society (France) or can be prepared from malonic acid and acetone.[3] The checkers purchased Meldrum's acid from the Aldrich Chemical Company, Inc.

3. Nickel acetylacetonate monohydrate is a better basic catalyst than triethylamine for the condensation of Meldrum's acid and the lactim ether. The yields are higher and the product is easier to purify.

4. The submitters report mp 145-147°C.

5. Absolute ethanol must be freshly distilled to obtain good yields in the transesterification.

6. The submitters report mp 48-50°C. The product, ethyl α-(hexahydroazepinylidene-2)acetate, shows a Z geometry. The ^{1}H NMR (300 MHz) spectrum of this compound is as follows: δ: 1.22 (t, 3 H, J = 7.1), 1.65 (m, 6 H), 2.25 (m, 2 H), 3.25 (m, 2 H), 4.06 (q, 2 H, J = 7.1), 4.42 (s, 1 H), 8.83 (br s, 1 H).

3. Discussion

This procedure is representative of a general and versatile method for the preparation of cyclic β-enamino esters which are known to be precursors of many alkaloids such as camptothecin,[4] (±)-lamprolobine,[5] (±)-lupinine[6] or isoretronecanol.[7]

Common synthetic methods for the preparation of cyclic β-enamino esters are the condensation between a lactim ether and benzyl cyanoacetate followed by hydrogenolytic decarboxylation,[8] or the imino ester carbon-carbon condensation with tert-butyl cyanoacetate followed by a trifluoroacetic acid treatment.[9] The use of a thiolactam condensed with ethyl bromoacetate gives, after sulfur extrusion by triphenylphosphine,[10] cyclic β-enamino esters. Compared with these methods, the Meldrum's acid condensation followed by the monodecarboxylating transesterification described here is more convenient and practical. An extension of this procedure permits preparation of smaller cyclic β-enamino esters in comparable yields.[11] The results are reported in the Table below.

TABLE

PREPARATION OF SMALL-RING β-ENAMINO ESTERS

Product	n	Yield	mp (solvent) or bp/mm	References
A	3	92-94%	170-172°C (ethanol)	11
B	3	85-87%	61-63°C (hexane)	2,7,11
A	4	90-92%	118-120°C (ethanol)	11
B	4	80-84%	90-94°C/0.1 mm	2,5,6,11

Only ethyl or methyl esters can be prepared by this procedure. However, pyrolysis of the cyclic β-enamino diesters at 225°C in the presence of different alcohols, thiols, or amines is a versatile and rapid method for preparing cyclic β-enamino esters, thioesters or amides.[2]

1. Laboratoire de Chimie des Hétérocycles and U.A. 455. Université P et M. Curie, Paris, France.
2. Celerier, J.-P.; Lhommet, G.; Maitte, P. *Tetrahedron Lett.* **1981**, *22*, 963.
3. Benson, R. E.; Cairns, T. L. *Org. Synth., Collect. Vol. IV* **1963**, 588.
4. Danishefsky, S.; Etheredge, S. J. *J. Org. Chem.* **1974**, *39*, 3430.
5. Yamada, Y.; Hatano, K.; Matsui, M. *Agr. Biol. Chem.* **1970**, *34*, 1536.
6. Gerrans, G. C.; Howard, A. S.; Orlek, B. S. *Tetrahedron Lett.* **1975**, 4171.
7. Pinnick, H. W.; Chang, Y.-H. *J. Org. Chem.* **1978**, *43*, 4662.
8. Oishi, T.; Nagai, M.; Onuma, T.; Moriyama, H.; Tsutae, K.; Ochiai, M.; Ban, Y. *Chem. Pharm. Bull.* **1969**, *17*, 2306.
9. Bertele, E.; Boos, H.; Dunitz, J. D.; Elsinger, F.; Eschenmoser, A.; Felner, I.; Gribi, H. P.; Gschwend, H.; Meyers, E. F.; Pesaro, M.; Scheffold, R. *Angew. Chem.* **1964**, *76*, 393; *Angew. Chem. Intern. Ed. Engl.* **1964**, *3*, 490.
10. Yamada, Y.; Miljkovic, D.; Wehrli, P.; Golding, B.; Löliger, P.; Keese, R.; Müller, K.; Eschenmoser, A. *Angew. Chem.* **1969**, *81*, 301; *Angew. Chem., Intern. Ed. Engl.* **1969**, *8*, 343.
11. Celerier, J.-P.; Deloisy, E.; Lhommet, G.; Maitte, P. *J. Org. Chem.* **1979**, *44*, 3089.

Appendix

Chemical Abstracts Nomenclature (Collective Index Number);

(Registry Number)

Ethyl α-(hexahydroazepinylidene-2)acetate: Acetic acid, (hexahydro-2H-azepin-2-ylidene)-, ethyl ester, (Z)- (10); (70912-51-5)

Isopropylidene α-hexahydroazepinylidene-2)malonate: 1,3-Dioxane-4,6-dione, 5-(hexahydro-2H-azepin-2-ylidene)-2,2-dimethyl- (10); (70912-54-8)

O-Methylcaprolactim: 2H-Azepine, 3,4,5,6-tetrahydro-7-methoxy- (8,9); (2525-16-8)

Meldrum's Acid: 2,2-Dimethyl-1,3-dioxane-4,6-dione: Malonic acid, cyclic isopropylidene ester (8); 1,3-Dioxane-4,6-dione, 2,2-dimethyl- (9); (2033-24-1)

Nickel acetylacetonate: Nickel, bis(2,4-pentanedionato)- (8); Nickel, bis(2,4-pentanedionato-O,O')-, (SP-4-1) (9); (3264-82-2)

SELECTIVE CYCLOPROPANATION OF (S)-(-)-PERILLYL ALCOHOL: 1-HYDROXYMETHYL-4-(1-METHYLCYCLOPROPYL)-1-CYCLOHEXENE

(1-Cyclohexene-1-methanol, 4-(1-methylcyclopropyl)-)

Submitted by Keiji Maruoka, Soichi Sakane, and Hisashi Yamamoto.[1]
Checked by Hisatoyo Kato and Ryoji Noyori.

1. Procedure

A dry, 1-L, three-necked, round-bottomed flask is equipped with a gas inlet, 50-mL pressure-equalizing dropping funnel, rubber septum, and a Teflon-coated magnetic stirring bar. The flask is flushed with argon, after which 10.65 g (0.07 mol) of (S)-(-)-perillyl alcohol (Note 1) followed by 350 mL of dichloromethane (Note 2) is injected through the septum into the flask. The solution is stirred and 37.3 mL (0.147 mol) of triisobutylaluminum (Note 3) is added from the dropping funnel over a period of 20 min at room temperature (Note 4). After the mixture is stirred at room temperature for 20 min, 7.3 mL (0.091 mol) of diiodomethane (Note 5) is added dropwise with a syringe over a 10-min period. The mixture is stirred at room temperature for 4 hr, and poured into 400 mL of ice-cold 8% aqueous sodium hydroxide. The organic layer is separated, and the aqueous layer is extracted twice with 100-mL portions of dichloromethane. The combined extracts are dried over anhydrous sodium sulfate, and concentrated with a rotary evaporator at ca. 20 mm. The residual oil is distilled under reduced pressure to give 10.64-11.13 g (92-96%) of 1-hydroxymethyl-4-(1-methylcyclopropyl)-1-cyclohexene as a colorless liquid, bp 132-134°C (24 mm) (Notes 6 and 7).

2. Notes

1. (S)-(-)-Perillyl alcohol is available from Aldrich Chemical Company, Inc.

2. Reagent-grade dichloromethane was dried and stored over Linde type 4 A molecular sieves.

3. Neat triisobutylaluminum of 97.6% purity was supplied in a metal cylinder from Toyo Stauffer Chemical Company, Ltd. (Japan). This reagent is also available from Aldrich Chemical Company, Inc. Since neat triisobutylaluminum is pyrophoric and reacts violently with oxygen and water, the used syringe should be immediately washed with hexane.

4. During this operation an exothermic reaction took place.

5. Diiodomethane, available from Tokyo Kasei Kogyo Company, Ltd. (Japan), was used without any purification.

6. The spectral properties of the product are as follows: ^{1}H NMR ($CDCl_3$, 500 MHz) δ: 0.22 and 0.26 (m, 4 H, cyclopropyl C-H), 0.80-0.92, 1.24-1.30, and 1.36-1.47 (m, 3 H, cyclohexenyl C-H), 0.93 (s, 3 H, CH_3), 1.77-1.83 (m, 1 H, cyclohexenyl =C-C-H), 1.91-2.16 (m, 4 H, OH and cyclohexenyl =C-C-H), 3.99 (br t, 2 H, CH_2-O), 5.69 (br s, 1 H, =C-C-H); IR (liquid film) cm^{-1}: 3330, 2830-2960, 1423-1460, 1390, 1010, 1000.

7. Gas chromatographic analysis of the trimethylsilyl ether using a 25-m PEG-HT capillary column at 100°C indicated a purity of 93% (retention time: 11.2 min). Under the present conditions, neither the starting perillyl alcohol nor the isomeric monocyclopropanation product (1-hydroxymethyl-4-isopropenylbicyclo[4.1.0]heptane) were detected. Dicyclopropanation products amounted to less than 5%.

Discussion

This procedure illustrates a new method for selective cyclopropanation of unsaturated alcohols not obtainable with ordinary cyclopropanation reactions.[2] The selectivity in this trialkylaluminum-promoted cyclopropanation is complementary to that obtained in the Simmons-Smith reaction and its modifications,[3] which give facile hydroxyl-assisted cyclopropanations with perillyl alcohol to afford 1-hydroxymethyl-4-isopropenylbicyclo[4.1.0]heptane predominantly. A similar tendency was observed in the case of geraniol. Thus, cyclopropanation with the i-Bu_3Al/CH_2I_2 system takes place almost exclusively at the C(6)-C(7) olefinic site far from the hydroxyl group of geraniol, and the C(2)-C(3) olefinic bond is left intact.[2]

The present cyclopropanation using trialkylaluminum-methylene iodide may proceed via dialkyl(iodomethyl)aluminum as an active intermediate,[4] which can be also generated by the reaction of dialkylaluminum iodide with diazomethane.[5] In addition, reaction of diiodomethane with triisobutyl-aluminum (each 1 equiv) afforded nearly 1 equiv of isobutyl iodide as a product, suggesting the formation of diisobutyl(iodomethyl)aluminum in the solution.[2]

The combined use of a wide variety of trialkylaluminum compounds and alkylidene iodide serves as a highly convenient and versatile method for cyclopropanation of simple olefins under mild conditions.[2] For example, treatment of 1-dodecene with CH_2I_2/R_3Al (R = Me, Et, i-Bu) in dichloromethane at room temperature for 3-8 hr gave decylcyclopropane in 96-98% yields.

1. Department of Applied Chemistry, Faculty of Engineering, Nagoya University, Chikusa, Nagoya 464, Japan.
2. Maruoka, K.; Fukutani, Y.; Yamamoto, H. *J. Org. Chem.* **1985**, *50*, 4412.
3. Simmons, H. E.; Cairns, T. L.; Vladuchick, S. A.; Hoiness, C. M. *Org. React.* **1973**, *20*, 1.
4. Miller, D. B. *Tetrahedron Lett.* **1964**, 989.
5. Hoberg, H. *Justus Liebigs Ann. Chem.* **1962**, *656*, 1.

Appendix
Chemical Abstracts Nomenclature (Collective Index Number);
(Registry Number)

(S)-(-)-Perillyl alcohol: 1-Cyclohexene-1-methanol, 4-(1-methylethenyl)- (8,9); (536-59-4)

1-Hydroxymethyl-4-(1-methylcyclopropyl)-1-cyclohexene: 1-Cyclohexene-1-methanol, 4-(1-methylcyclopropyl)- (11); (98678-72-9)

3'-NITRO-1-PHENYLETHANOL BY ADDITION OF METHYLTRIISOPROPOXY-TITANIUM TO m-NITROBENZALDEHYDE

(Benzenemethanol, α-methyl-3-nitro-)

Submitted by René Imwinkelried and Dieter Seebach.[1]
Checked by Cheryl A. Martin and K. Barry Sharpless.

1. Procedure

A dry, 500-mL, three-necked flask equipped with a pressure-equalizing 100-mL dropping funnel, argon inlet, and magnetic stirrer is evacuated and flushed with argon (3 cycles). The flask is charged with 16.0 mL (57.7 mmol) of tetraisopropyl orthotitanate (Note 1) via a plastic syringe and hypodermic needle and 2.1 mL (19.2 mmol) of titanium tetrachloride is added over 5 min, with gentle cooling of the flask in an ice-water bath, to give a viscous oil (Note 2). After the addition of 70 mL of tetrahydrofuran (Note 3), the clear solution is stirred at room temperature for 30 min. The dropping funnel is charged with 62 mL (77 mmol, 1.24 M in hexane) of methyllithium (Note 4), which is added to the cooled (ice bath) tetrahydrofuran solution over a period of 25-30 min. During the addition the resulting suspension changes from orange to bright yellow. After the mixture has stirred at ice-bath temperature for 1 hr, a solution of 10.6 g (70 mmol) 3-nitrobenzaldehyde (Note 5) in 60 mL of tetrahydrofuran (Note 3) is added from the dropping funnel within 20-

25 min at the same temperature. The mixture is stirred at 0-5°C for 1 hr and then 60 mL of 2 N hydrochloric acid is added. The organic phase is separated in a separatory funnel and the aqueous phase is extracted with three 150-mL portions of diethyl ether. The combined organic phases are washed with 100 mL of saturated sodium bicarbonate solution and 100 mL of saturated sodium chloride solution and then dried over anhydrous magnesium sulfate. After filtration the solution is concentrated on a rotary evaporator and dried at 0.1 mm for 1 hr. The residue, 11.0-11.1 g (94-95%) of an orange-brown viscous oil, sometimes solidifies on standing (mp 55-60°C); the purity of the crude product is at least 95% (estimated by ^{1}H NMR). The product can be purified by short-path distillation at 120-125°C (0.15 mm) to give 9.9-10.4 g (85-89%) of a yellow oil, which solidifies on standing at room temperature or at -30°C in a freezer, mp 60.5-62.0°C (lit.[2] mp 62°C) (Note 6).

2. Notes

1. Commercial tetraisopropyl orthotitanate [Ti(O-i-Pr)$_4$] (Dynamit Nobel) and titanium tetrachloride (Fluka pract.) can be used without further purification. The checkers obtained Ti(O-i-Pr)$_4$ from Aldrich Chemical Company, Inc. and titanium tetrachloride from Fluka. Distillation of Ti(O-i-Pr)$_4$ did not improve the results.

2. If the mixture is overcooled, the resulting chlorotriisopropoxytitanium partially solidifies.

3. Tetrahydrofuran was distilled from potassium/benzophenone immediately before use.

4. The methyllithium solution was obtained from Metallgesellschaft, Frankfurt. The checkers used methyllithium (Aldrich Chemical Company, Inc.), salt free 1.4 M in ethyl ether, with no significant difference observed in the reaction.

5. 3-Nitrobenzaldehyde is Fluka purum, used without further purification. The checkers obtained it from Aldrich Chemical Company, Inc.

6. The product obtained after distillation can be recrystallized from benzene/petroleum ether (3:2, v:v) to give pale yellow crystals (91-95% from distilled product) with a melting point of 61-63°C. Attempts by the checkers to crystallize the crude reaction mixture were unsuccessful. ^{1}H NMR (CDCl$_3$) δ: 1.54 (d, 3 H, J = 6.5, CH$_3$), 2.3 (br, 1 H, OH), 5.00 (q, J = 6.5, O-C-H), 7.5-7.7 (m, 2 H, arom. H), 8.0-8.25 (m, 2 H, arom. H); IR (KBr), cm^{-1}: 3260 (br, m), 2990 (m), 1580 (m), 1525 (s), 1340 (s), 1205 (m), 1170 (m), 810 (m), 740 (m), 690 (m).

3. Discussion

The addition of nucleophilic organometallic compounds (usually RLi or RMgX) to a carbonyl group - a key step in numerous syntheses - is not always straightforward. The addition reaction is complicated by the fact that aldehydes, ketones, and esters are not well differentiated, that other electrophilic functional groups such as cyano, nitro, halo, trialkylstannyl may interfere, and that proton abstraction or one electron-transfer processes rather than addition occur. For example, the addition of methyllithium or methylmagnesium iodide to 3-nitrobenzaldehyde under the same conditions used with CH$_3$Ti(OCH(CH$_3$)$_2$)$_3$ (this procedure) leads to a complex mixture of products with formation of only 10-30% of 3'-nitro-1-phenylethanol.[3] In many cases these complications can be remedied by using derivatives of titanium and zirconium, compounds which have become increasingly important in organic syntheses during the past decade. Several review articles discuss different aspects.[3]

The nucleophilic titanium and zirconium reagents are readily available by simple transmetallation of the organolithium or Grignard reagents with $(RO)_3TiCl$, $(RO)_3ZrCl$, or $(R_2N)_3TiX$. The trialkoxychloro compounds are prepared by mixing the inexpensive, industrially available, titanates, $Ti(OR)_4$, or zirconates, $Zr(OR)_4$, with the appropriate amount of $TiCl_4$ or $ZrCl_4$. In contrast to compounds of most other heavy metals, few toxic effects of $Ti(OR)_4$ and $Zr(OR)_4$ are known, partly because they are very rapidly hydrolyzed by water and the resulting oxide-hydrates are insoluble (TiO_2 is a white pigment). Some of the reagents, $RTi(OR')_3$, can be isolated without difficulty. Thus, $CH_3Ti(O-i-Pr)_3$ can be obtained as a bright yellow oil which distills without decomposition at 50°C/0.001 mm.[4]

The organo-titanium and -zirconium compounds, for the most part generated in situ, react highly selectively with carbonyl compounds. For example, $CH_3Ti(O-i-Pr)_3$ reacts five orders of magnitude faster with benzaldehyde than with acetophenone at room temperature.[5] Reagents of the type $RTi(O-i-Pr)_3$ add smoothly to nitro- (see this procedure), iodo-, or cyano-substituted benzaldehydes, and the reactions may be performed in chlorinated solvents or in acetonitrile (for some examples see Table). The zirconium analogues have particularly low basicity and add in high yield to α- and β-tetralones.[6] The inclusion of chiral OR* groups gives enantioselective reagents (up to $\geq 98\%$ ee).[7,8,9] Allylic $(RO)_3Ti$-derivatives react diastereoselectively only at the more highly substituted carbon atom with aldehydes and even with unsymmetrical ketones.^{8,9,10} Titanates can be used as mild catalysts for the transesterification of compounds containing acid- or base-labile functional groups.^{11}

1. Laboratorium für Organische Chemie der Eidgenössischen Technischen Hochschule, ETH-Zentrum, Universitätstrasse 16, CH-8092 Zürich, Switzerland.

2. Arcus, C. L.; Schaffer, R. E. *J. Chem. Soc.* **1958**, 2428.

3. For a review with practical and experimental aspects see: Seebach, D.; Weidmann, B.; Widler, L. In "Modern Synthetic Methods 1983: Transition Metals in Organic Synthesis"; Scheffold, R., Ed.; Wiley: New York and Salle + Sauerländer, Aarau, Switzerland, 1983. Further reviews: Weidmann, B.; Seebach, D. *Angew. Chem.* **1983**, *95*, 12; *Angew. Chem. Intern. Ed. Engl.* **1983**, *22*, 31; Reetz, M. T. *Top. Curr. Chem.* **1982**, *106*, 1.

4. Clauss, K. *Justus Liebigs Ann. Chem.* **1968**, *711*, 19.

5. Weidmann, B.; Seebach, D. *Helv. Chim. Acta* **1980**, *63*, 2451.

6. Weidmann, B.; Maycock, C. D.; Seebach, D. *Helv. Chim. Acta* **1981**, *64*, 1552.

7. Seebach, D.; Beck, A. K.; Roggo, S.; Wonnacott, A. *Chem. Ber.* **1985**, *118*, 3673.

8. Seebach, D. In "Organic Synthesis, an Interdisciplinary Challenge", Proc. 5th IUPAC Symp. Org. Synth., Streith, J.; Prinzbach, H.; Schill, G., Eds., Blackwell Scientific Publications: Oxford, 1985.

9. Seebach, D.; Beck, A. K.; Schiess, M.; Widler, L.; Wonnacott, A. *Pure Appl. Chem.* **1983**, *55*, 1807.

10. Widler, L.; Seebach, D. *Helv. Chim. Acta* **1982**, *65*, 1085; Seebach, D.; Widler, L.; *Helv. Chim. Acta* **1982**, *65*, 1972.

11. Imwinkelried, R.; Schiess, M.; Seebach, D. *Org. Synth.* **1987**, *65*, 230; Schnurrenberger, P.; Züger, M. F.; Seebach, D. *Helv. Chim. Acta* **1982**, *65*, 1197; Seebach, D.; Züger, M. *Helv. Chim. Acta* **1982**, *65*, 495; Seebach, D.; Hungerbühler, E.; Naef, R.; Schnurrenberger, P.; Weidmann, B.; Züger, M. *Synthesis* **1982**, 138; Rehwinkel, H.; Steglich, W. *Synthesis* **1982**, 826.

Appendix
Chemical Abstracts Nomenclature (Collective Index Number); (Registry Number)

3'-Nitro-1-phenylethanol: Benzyl alcohol, α-methyl-m-nitro- (8); Benzenemethanol, α-methyl-3-nitro- (9); (5400-78-2)

m-Nitrobenzaldehyde: Benzaldehyde, m-nitro- (8); Benzaldehyde, 3-nitro- (9); (99-61-6)

Methyltriisopropoxytitanium: Titanium, triisopropoxymethyl- (8); Titanium, methyltris(2-propanolato)-, (T-4)- (9); (18006-13-8)

Tetraisopropyl orthotitanate: Isopropyl alcohol, titanium (4+) salt (8); 2-Propanol, titanium (4+) salt (9); (546-68-9)

TABLE

SOME PRODUCTS OF ORGANOTITANIUM TRIISOPROPOXIDES [RTi(OiPr)$_3$] WITH FUNCTIONALIZED CARBONYL COMPOUNDS.[3] THE BONDS MADE DURING THE REACTION ARE DRAWN BOLD.

N-ACETYL-N-PHENYLHYDROXYLAMINE VIA CATALYTIC TRANSFER HYDROGENATION OF NITROBENZENE USING HYDRAZINE AND RHODIUM ON CARBON

(Acetamide, N-hydroxy-N-phenyl-)

$$\underset{\text{THF, 30°C}}{\overset{\text{5\% Rh-C, N}_2\text{H}_4 \cdot \text{H}_2\text{O}}{\longrightarrow}} \text{PhNO}_2 \longrightarrow \text{PhNHOH} \underset{\substack{\text{CH}_2\text{Cl}_2,\ \text{H}_2\text{O} \\ 0°\text{C}}}{\overset{\text{AcCl, NaHCO}_3}{\longrightarrow}} \text{Ph-N(Ac)-OH}$$

Submitted by P. W. Oxley, B. M. Adger, M. J. Sasse, and M. A. Forth.[1]
Checked by Soo Y. Ko and K. Barry Sharpless.

1. Procedure

Caution! Nitrobenzene and hydrazine are both toxic. Phenylhydroxylamine and N-acetyl-N-phenylhydroxylamine are both suspected carcinogens.

A. *N-Phenylhydroxylamine.* Wet, 5% rhodium on carbon (1.1 g) (Note 1), tetrahydrofuran (200 mL) (Note 2) and nitrobenzene (41.0 g) (Note 3) are introduced into a 500-mL, three-necked, round-bottomed flask fitted with a mechanical stirrer, thermometer and condenser. The mixture is cooled to 15°C and hydrazine hydrate (17.0 g) (Note 4) is introduced into the reaction mixture from a pressure-equalized addition funnel over 30 min. The temperature of the mixture is maintained at 25-30°C throughout the addition by means of an ice-water bath. After the mixture is stirred for a further 2 hr at 25-30°C, the reaction is complete (Note 5). The mixture is filtered and the catalyst washed with a little tetrahydrofuran. The solution is used immediately in the acylation step (Note 6).

B. *N-Acetyl-N-phenylhydroxylamine.* To the N-phenylhydroxylamine solution in a 1000-mL, three-necked, round-bottomed flask fitted with a mechanical stirrer and thermometer is added a slurry of sodium bicarbonate (42 g) in water (40 mL). The mixture is cooled to -4°C in an ice-salt bath before acetyl chloride (26.0 g) (Note 7) is introduced into the well-stirred mixture over 1 hr (Note 8) while the temperature is maintained below 0°C. Stirring is then continued for 30 min before a solution of sodium hydroxide (20.0 g) in water (200 mL) is added keeping the temperature below 20°C. The aqueous phase is separated, the tetrahydrofuran phase is diluted with an equal volume of petroleum ether, the aqueous phase is separated again, and the organic phase is extracted with aqueous 10% sodium hydroxide solution (2 x 50 mL). The combined aqueous phases are washed with methylene chloride (200 mL) and then neutralized with concentrated hydrochloric acid (cooling employed). The mixture is extracted with methylene chloride (3 x 100 mL) and the extracts are combined, dried over magnesium sulfate, filtered and concentrated at reduced pressure (about one fifth volume) (Note 9). After the solution is cooled to 40°C, 100 mL of petroleum ether (bp 60-80°C) is added. The mixture is stirred at 10°C for 30 min before filtering and washing with additional petroleum ether. The material is dried at room temperature to afford 39.3-40.1 g (79-80%) of N-acetyl-N-phenylhydroxylamine as a white crystalline solid, mp 66-67°C [lit.[2] mp 67-67.5°C] (Note 10).

2. Notes

1. The 5% rhodium on carbon used was purchased dry from Engelhard Industries Ltd. The checkers purchased it from Aldrich Chemical Company, Inc. The catalyst is used wet (40-50% water) to reduce the risk of fire when the solvent is added.

2. Tetrahydrofuran was from a bulk supply purchased from Blagden Campbell. The checkers obtained it from EM Science. The solvent was tested for peroxides prior to use.

3. Nitrobenzene was supplied by BDH Chemicals Ltd., and was used as received. The checkers obtained it from Aldrich Chemical Company, Inc. Nitrobenzene should be handled only with gloves and in an efficient fume hood.

4. Hydrazine hydrate was purchased from FBC Industrial Chemicals and was used as supplied. The checkers obtained it from Aldrich Chemical Company, Inc. Hydrazine is a severe poison and should be handled only with gloves in an efficient fume hood.

5. An HPLC system was used to monitor the reduction and to determine the end of the reaction. The HPLC monitoring was not employed by the checker. However, TLC indicated that the reduction was almost complete after stirring for 2 hr at 25-30°C. If only a slight excess (1.03 equiv) of hydrazine is employed the reaction is generally complete in 2 hr and excessive over-reduction cannot occur.

The HPLC system consisted of a Waters C_{18} μ-Bondapak column, a mobile phase consisting of 15% acetonitrile, 85% 0.05 M aqueous ammonium acetate using a flow rate of 2 mL/min, and a UV wavelength detector for 235 nm. The relative response factors of nitrobenzene and aniline were 1.75 and 0.66, respectively.

6. N-Phenylhydroxylamine, mp 83.5-85°C, can be isolated at this stage in 75-85% yields if desired, but it should be borne in mind that N-phenylhydroxylamine is not very stable. The isolation can be carried out by adding an equal volume of methylene chloride to the tetrahydrofuran solution which is then dried over magnesium sulfate and concentrated to low volume under reduced pressure. Addition of a little petroleum ether precipitates N-phenylhydroxylamine which is then filtered and washed with petroleum ether.

7. Acetyl chloride was obtained from Hoechst and was used as supplied. The checkers obtained it from Fluka Chemical Corporation. The quantity of acetyl chloride used is 1.05 equiv based on the HPLC yield. (The checkers simply used the amount specified.) Acetyl chloride should be handled only with gloves in an efficient fume hood.

8. No vigorous, exothermic reaction is seen during the addition of acetyl chloride, but the additon should be slow because of the heterogeneous nature of the reaction and the need to destroy efficiently hydrogen chloride as it is formed. The product, like N-phenylhydroxylamine, is sensitive to acid and undergoes the Bamberger rearrangement.[3]

9. Excessive heating causes decomposition of the product. This method also affords an easily handled crystalline solid of good purity.

10. The following analytical data have been obtained: ^{1}H NMR (CDCl$_3$, 100 MHz) δ: 2.11 (s, 3 H acetylmethyl); 7.40 (m, 5 H, aromatics); 8.90 (broad, 0.6 H, NOH); IR (Nujol) cm^{-1}: 3140, 2930, 2860, 1630, 1595, 1460, 1380. Anal. Calcd. for $C_8H_9NO_2$: C, 63.56; H, 6.00; N, 9.27. Found C, 63.48; H, 5.99; N, 9.21. Non-aqueous titration (Bu$_4$NOH), 98.3%.

3. Discussion

This preparation illustrates a convenient reduction of nitrobenzene under catalytic transfer hydrogenation conditions to give N-phenylhydroxylamine in high yield and demonstrates a mono-acylation method to afford the N-acetyl derivative in high yield. Some work has been done in this area by Johnstone, et al.[4] A number of other reductive methods described in the literature were tried,[5,6,7] but these were not as good as the procedure described here. Phenylhydroxylamine is thermally unstable, can undergo a Bamberger rearrangement,[3] and deteriorates on storage, so its isolation is undesirable. The material was therefore converted directly, without isolation, to its more stable N-acetyl derivative. Other acylation methods led to mixtures of mono and diacylated products.

1. Smith Kline and French Research Ltd., Old Powder Mills, Leigh, Tonbridge, Kent, TN11 9AN, England.
2. Dictionary of Organic Compounds, 5th Ed.; Chapman and Hall: New York; 1982.
3. For reviews: See Shine, H. J. "Aromatic Rearrangements", Elsevier: New York, 1967; pp 182-190; Hughes, E. D.; Ingold, C. K. *Q. Rev., Chem. Soc.* **1952**, *6*, 34-62 (especially pp 45-48).
4. Entwistle, I. D.; Gilkerson, T.; Johnstone, R. A. W. Telford; R. P. *Tetrahedron* **1978**, *34*, 213; Brit. Pat. 1 575 808.
5. Kamm, O. *Org. Synth., Collect. Vol. I* **1941**, 445.
6. Ger. Offen. 2 455 238; *Chem. Abstr.* **1975**, *83*, 147284t; Ger. Offen. 2 455 887; *Chem. Abstr.* **1975**, *83*, 113918n; Ger. Offen. 2 327 412; *Chem. Abstr.* **1974**, *80*, 70516g.
7. Ger. Offen. 2 118 369; *Chem. Abstr.* **1972**, *76*, 14082j.

Appendix

Chemical Abstracts Nomenclature (Collective Index Number); (Registry Number)

N-Phenylhydroxylamine: Hydroxylamine, N-phenyl- (8); Benzenamine, N-hydroxy- (9); (100-65-2)

N-Acetyl-N-phenylhydroxylamine: Acetohydroxamic acid, N-phenyl- (8); Acetamide, N-hydroxyl-N-phenyl- (9); (1795-83-1)

METHYL 7-HYDROXYHEPT-5-YNOATE

(5-Heptynoic acid, 7-hydroxy-, methyl ester)

A. $BrCH_2CH_2CH_2CN \xrightarrow{\text{CH}_3\text{OH}}_{\text{HCl}} BrCH_2CH_2CH_2C(=NH_2^+Cl^-)OCH_3$

B. $BrCH_2CH_2CH_2C(=NH_2^+Cl^-)OCH_3 \xrightarrow{\text{CH}_3\text{OH}}_{\text{hexane}} BrCH_2CH_2CH_2C(OCH_3)_3$

C. $HOCH_2\equiv CH \xrightarrow{\text{2 LiNH}_2}_{\text{NH}_3 \text{[liq]}} LiOCH_2C\equiv CLi \xrightarrow{\text{(i) } BrCH_2CH_2CH_2C(OCH_3)_3}_{\text{(ii) } H_3O^+}$

$HOCH_2C\equiv CCH_2CH_2CH_2CO_2CH_3$

Submitted by Guy Casy,[1] John W. Patterson,[2] and Richard J. K. Taylor.[1]
Checked by Friedhelm Balkenhohl and E. Winterfeldt.

1. Procedure

A. Methyl 4-bromo-1-butanimidate hydrochloride. A 500-mL, three-necked, round-bottomed flask is connected via neck A to a hydrogen chloride water trap using the arrangement shown in Figure 1. A stream of nitrogen (Note 1) is introduced via neck C; the flask is flame-dried and allowed to cool. The stopper is removed from neck B and the flask is charged with 29.6 g (0.20 mol) of 4-bromobutanenitrile (Note 2), 200 mL of dry ether (Note 3) and 7.7 g (0.24 mol; 1.2 equiv based on 1.0 equiv of 4-bromobutanenitrile) of dry methanol (Note 3). The stopper is replaced, and the weight of the flask and its contents are recorded. The flask and its contents are cooled to -5°C by immersion in an ice-salt bath, the nitrogen source is removed, and the gas inlet tube is connected to the cylinder of hydrogen chloride. The cylinder tap is cautiously opened, and hydrogen chloride is allowed to bubble through the reaction mixture at a steady but controlled rate until 18.2 g (0.50 mol; 2.5 equiv based on 1.0 equiv of 4-bromobutanenitrile) has been absorbed (Note 4). A stopper is placed in each neck of the flask and a strip of Parafilm is bound around the edge of each ground-glass connnection to ensure an airtight seal.

Figure 1

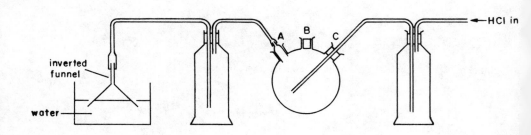

The flask is stored at 5°C (refrigerator) for 4-5 days (Note 5), after which time a copious precipitate of the title compound is obtained. The mixture is filtered with suction through a dry 100-mm sintered glass funnel. After all of the product is collected, a large inverted funnel connected to a nitrogen source is positioned about 15 cm above the sintered funnel to provide a blanket of dry, inert gas (Note 6). The product is washed thoroughly with several portions of dry ether, totaling 500 mL, and then dried to constant weight over solid potassium hydroxide in a desiccator evacuated at 12-20 mm (water aspirator). There is obtained 39.0 g (90%) (Note 7) of methyl 4-bromo-1-butanimidate hydrochloride as fine white crystals, mp 95-97°C (Note 8).

B. *Trimethyl ortho-4-bromobutanoate*. A 1-L, two-necked, round-bottomed flask containing an efficient magnetic stirring bar is purged with nitrogen (Note 1) introduced via a pressure-equalizing glass bubbler (Note 9). The flask is charged with 38.9 g (0.18 mol) of methyl 4-bromo-1-butanimidate hydrochloride, 450 mL of dry hexane (Note 3) and 17.3 g (0.54 mol) of dry methanol (Note 3). A stopper is placed in one neck of the flask; the glass bubbler is removed from the other neck and immediately replaced with a gas outlet adapter to which is attached a balloon filled with nitrogen (Note 10). Finally, a strip of Parafilm is bound around the edge of each ground-glass connection to ensure an airtight seal. The reaction mixture is stirred at room temperature for 48 hr, then filtered with suction to remove the precipitated ammonium chloride. The filter cake is washed with two 30-mL portions of dry hexane, and the filtrate and washings are concentrated under reduced pressure (water aspirator) at 30-40°C by rotary evaporation to leave a slightly turbid, colorless liquid, to which is added 0.25 g of anhydrous potassium carbonate. This material is distilled under reduced pressure to afford 36.4-36.7 g (89-90%) (Note 11) of trimethyl ortho-4-bromobutanoate as a colorless oil, bp 65°C (0.5 mm) (Note 12).

C. *Methyl 7-hydroxyhept-5-ynoate*. A 2-L, three-necked, round-bottomed flask equipped with a dry ice condenser and a mechanical stirring rod is charged with 750 mL of anhydrous ammonia, via a gas-inlet tube, at -33°C (Note 13) under nitrogen (Notes 1 and 14). The gas-inlet tube is removed, and about 0.1 g of lithium wire (Note 2) is added in small portions until a permanent blue color is obtained. Ferric nitrate (0.1 g) is added to discharge the blue color, and after the solution is stirred for 5 min, 4.24 g (0.611 mol; 2.5 equiv based on 1.0 equiv of propargyl alcohol) of lithium wire is added in small portions. After the addition is complete the flask is fitted with a 100-mL pressure-equalizing and serum-capped dropping funnel. Stirring is continued for 20 min to obtain a grey suspension of lithium amide, to which is added dropwise a solution of 13.7 g (0.245 mol; 1.5 equiv based on 1.0 equiv of trimethyl ortho-4-bromobutanoate) of redistilled propargyl alcohol (Note 2) in 15 mL of dry ether (Note 3). After the solution is stirred for 20 min, a solution of 36.4 g (0.160 mol) of trimethyl ortho-4-bromobutanoate in 40 mL of dry ether is added dropwise. Stirring is continued for 3 hr, the reaction vessel is opened to the atmosphere, and its contents are allowed to warm to room temperature over 16-18 hr. The mixture is heated at 50°C on a water bath under a stream of nitrogen to remove any remaining ammonia. This furnishes a grey solid, which is cooled to 0°C, and 5% sulfuric acid is added in 100-mL portions until a pH of 1 is obtained (Note 15). The resulting suspension is stirred at room temperature for 30 min, and extracted with three 200-mL portions of ether. The combined organic extracts are washed with 200 mL of saturated sodium bicarbonate, dried over magnesium sulfate, and filtered. The filtrate is concentrated under reduced pressure (water aspirator) at 30-40°C by rotary evaporation, to leave 19.1 g (77%) of an essentially pure amber oil (Note 16). This material can be distilled under reduced pressure to afford 16.8 g (67%) (Note 17) of methyl 7-hydroxyhept-5-ynoate as a colorless oil, bp 100°C (0.05 mm) (Notes 18 and 19).

2. Notes

1. Oxygen-free nitrogen, dried by passage through activated molecular sieves, was used.

2. 4-Bromobutanenitrile was obtained from Lancaster Synthesis Ltd. Alternatively it can be prepared from 1,3-dibromopropane and potassium cyanide using the procedure of Derrick and Henry.[3] Lithium wire (3.2 mm diameter, containing ca. 0.01% sodium) and propargyl alcohol were obtained from the Aldrich Chemical Company, Inc. The latter was dried with potassium carbonate and then distilled prior to use.

3. Diethyl ether and hexane were freshly distilled from blue solutions obtained with sodium and benzophenone. Methanol was distilled from magnesium and iodine. The use of high purity solvents (e.g., Mallinckrodt AR anhydrous ether, Nanograde hexane and AR methanol) as received gave only small reductions in yield.

4. To monitor uptake of hydrogen chloride the flask and its contents are periodically weighed. Typically, the process is complete within 5 min.

5. The progress of the reaction can be monitored by tilting the flask slightly to expose clear supernatant liquor. The flask is left in this position overnight, and if no further crystallization is apparent, the product may be isolated.

6. Alternatively a dry nitrogen glove box can be employed.

7. In several smaller scale experiments (0.07-0.14 mol of 4-bromobutanenitrile) yields in the range 83-93% were achieved.

8. The product has the following spectroscopic properties: IR (Nujol) cm^{-1}: 1650, 1405, 1215, 875; ^{1}H NMR (TFA-d) δ: 2.08-2.64 (m, 2 H), 3.04 (t, 2 H, J = 7), 3.48 (t, 2 H, J = 6), 4.32 (s, 3 H), 9.52 (br s, 2 H); ^{13}C NMR (TFA-d) δ: 28.53, 30.82, 33.41, 60.77, 183.65.

9. The glassware was dried overnight at 150°C and assembled hot under nitrogen.

10. This arrangement is preferable to a continuous flow of nitrogen; otherwise the quantity of methanol in situ may be critically diminished.

11. In several smaller scale experiments (0.05-0.12 mol of methyl 4-bromo-1-butanimidate hydrochloride) yields in the range 90-93% were achieved.

12. The product has the following spectroscopic properties: IR (neat) cm^{-1}: 2840, 1740, 1070; ^{1}H NMR (CDCl$_3$) δ: 1.77-2.03 (m, 4 H), 3.23 (s, 9 H), 3.33-3.60 (m, 2 H); ^{13}C NMR (CDCl$_3$) δ: 26.54, 29.00, 33.87, 49.37, 115.31.

13. To minimize evaporation of liquid ammonia, a temperature of -33 ± 5°C was maintained until the workup by means of a dry ice-acetone cooling bath. In addition, the dry ice condenser was continually charged with a saturated dry ice-acetone mixture.

14. The apparatus is maintained under a slight positive nitrogen pressure until the work up. If this precaution is not taken, atmospheric moisture is drawn into the apparatus (see Note 19).

15. A total volume of 500-600 mL of 5% sulfuric acid is normally required.

16. This material is essentially pure as indicated by TLC and ^{1}H NMR spectroscopic analyses, and by CH microanalysis. (Anal. Calcd for $C_8H_{12}O_3$: C, 61.5; H, 7.7. Found: C, 61.8; H, 7.9.)

17. On a smaller scale (0.109 mol of trimethyl ortho-4-bromobutanoate) in which a magnetic stirring bar was used to agitate the reaction mixture, a distilled yield of 71% (81% before distillation) was achieved.

18. The submitters report bp 120°C (0.06 mm). The product has the following spectroscopic properties: IR (neat) cm^{-1}: 3420, 1735, 1015; ^{1}H NMR (CDCl$_3$) δ: 1.63-2.63 (m, 6 H), 3.24 (s, 1 H), 3.67 (s, 3 H), 4.12-4.30 (m, 2 H); ^{13}C NMR (CDCl$_3$) δ: 18.02, 23.54, 32.64, 50.73, 51.43, 79.32, 84.31, 173.60.

19. If moisture is drawn into the apparatus, this can diminish the quantity of the dilithio derivative of propargyl alcohol. If any O-lithio monoanion is present during the addition of trimethyl ortho-4-bromobutanoate, a quantity of the O-alkylated derivative, methyl 4-(2-propynyloxy)butanoate will be produced. The latter exhibits the following ^{1}H NMR (CDCl$_3$) spectrum: δ: 1.61-2.58 (m, 5 H), 3.52 (t, 2 H, J = 5.5), 3.65 (s, 3 H), 4.09 (d, 2 H, J = 2.5).

3. Discussion

Methyl 7-hydroxyhept-5-ynoate is an important precursor to alkylating agents that are used to introduce the complete prostaglandin α-side chain.[4,5] It is normally prepared from propargyl alcohol using a six-step sequence originally introduced by Corey and Sachdev[6] with subsequent modifications.[7-11] Alternative routes to methyl 7-hydroxyhept-5-ynoate have also been reported[12,13] but appear less efficient than the one described here. The present route arose from the observation that whereas alkylations of propargyl alcohol-derived anions with 4-halobutanoates were unsuccessful,[13,14] the use of trimethyl ortho-4-bromobutanoate gave efficient alkylation.[14] Related reactions using orthoester protecting groups have been reported recently[15] and the preparation of such compounds from nitriles using the Pinner reaction, as described herein, is well established.[16]

1. School of Chemical Sciences, University of East Anglia, Norwich, NR4 7TJ, UK.
2. Syntex Research, 3401 Hillview Avenue, P.O. Box 10850, Palo Alto, CA 94304; Contribution No. 730 from the Institute of Organic Chemistry.
3. Derrick, C. G.; Henry, R. W. *J. Am. Chem. Soc.* **1918**, *40*, 537-558 (see p. 546).
4. "New Synthetic Routes to Prostaglandins and Thromboxanes"; Roberts, S. M.; Scheinmann, F., Eds.; Academic Press: London, 1982; Newton, R. F.; Roberts, S. M.; Taylor, R. J. K. *Synthesis* **1984**, 449.
5. Noyori, R.; Suzuki, M. *Angew. Chem., Intern. Ed. Engl.* **1984**, *23*, 847; Suzuki, M.; Yanagisawa, A.; Noyori, R. *J. Am. Chem. Soc.* **1985**, *107*, 3349 and references therein.
6. Corey, E. J.; Sachdev, H. S. *J. Am. Chem. Soc.* **1973**, *95*, 8483; see also Bagli, J.; Bogri, T. *Tetrahedron Lett.* **1972**, 3815.
7. Noguez, J. A.; Maldonado, L. A. *Synth. Commun.* **1976**, *6*, 39.
8. Patterson, J. W., Jr.; Fried, J. H. *J. Org. Chem.* **1974**, *39*, 2506.
9. Martel, J.; Blade-Font, A.; Marie, C.; Vivat, M.; Toromanoff, E.; Buendia, J. *Bull. Soc. Chim. Fr.* **1978**, (3-4, Pt. 2), 131.
10. Elder, J. S.; Mann, J.; Walsh, E. B. *Tetrahedron* **1985**, *41*, 3117; Luo, F.-T.; Negishi, E.-i. *J. Org. Chem.* **1985**, *50*, 4762.
11. Donaldson, R. E.; Saddler, J. C.; Byrn, S.; McKenzie, A. T.; Fuchs, P. L. *J. Org. Chem.* **1983**, *83*, 2167.
12. Ferdinandi, E. S.; Just, G. *Can. J. Chem.* **1971**, *49*, 1070.
13. Haynes, R. K. University of Sydney (Personal Communication).
14. Casy, G.; Furber, M.; Richardson, K. A.; Stephenson, G. R. and Taylor, R. J. K. *Tetrahedron* **1986**, *42*, 5849.

15. Patterson, J. W. *Synthesis* **1985**, 337.

16. Roger, R.; Neilson, D. G. *Chem. Rev.* **1961**, *61*, 179.

Appendix
Chemical Abstracts Nomenclature (Collective Index Number); (Registry Number)

Methyl 7-hydroxyhept-5-ynoate: 5-Heptynoic acid, 7-hydroxy-, methyl ester (9); (50781-91-4)

4-Bromobutanenitrile: Butyronitrile, 4-bromo- (8); Butanenitrile, 4-bromo- (9); (5332-06-9)

Trimethyl ortho-4-bromobutanoate: Butane, 4-bromo-, 1,1,1-trimethoxy- (9); (55444-67-2)

4-METHOXY-3-PENTEN-2-ONE

(3-Penten-2-one, 4-methoxy-)

$$\text{CH}_3\text{COCH}_2\text{COCH}_3 + \text{HC(OCH}_3)_3 \xrightarrow[\substack{\text{CH}_3\text{OH} \\ 55°\text{C/5 hr}}]{\text{C}_7\text{H}_7\text{SO}_3\text{H}} \underset{\text{CH}_3\text{O}}{\overset{\text{CH}_3}{>}}\text{C}=\text{CHCCH}_3$$

Submitted by George A. Kraus, Michael E. Krolski and James Sy.[1]
Checked by Yun Gao and K. Barry Sharpless.

1. Procedure

4-Methoxy-3-penten-2-one. A flame-dried, 250-mL, one-necked flask equipped with a condenser and drying tube is charged with 2,4-pentanedione (Note 1) (25.0 g, 250 mmol), trimethyl orthoformate (Note 2) (26.53 g, 250 mmol), p-toluenesulfonic acid (0.54 g, 2.8 mmol) and methanol (Note 3) (62 mL). The flask is placed in an oil bath and heated at 55°C for 5 hr. The solution is cooled and concentrated under reduced pressure. Fifty milliliters of CCl_4 is added and the solution is again concentrated under reduced pressure. The crude product is distilled via a short path condenser and collected in a flask cooled in an ice bath (Note 4). The product distills at 43-47°C (4 mm) at an oil bath temperature of 60°C (Note 5). The yield of pure product is 17.3-18.8 g (61-66%) (Note 6).

2. Notes

1. 2,4-Pentanedione was obtained from Aldrich Chemical Company, Inc. Its purity was greater than 99% and was used without purification.

2. The trimethyl orthoformate used in this experiment was obtained from Aldrich Chemical Company, Inc. Its purity was listed as 98% and was used without purification.

3. Methanol was obtained from Fisher Scientific. It was anhydrous grade methanol.

4. The checkers used a dry ice-acetone cooling bath.

5. Use of higher temperature (>65°C) results in a low yield.

6. The spectral properties of 4-methoxy-3-penten-1-one are as follows: IR (neat) cm^{-1}: 1674, 1590, 1165, 922. NMR (CDCl$_3$) δ: 2.15 (s, 3 H), 2.28 (s, 3 H), 3.64 (s, 3 H), 5.41 (s, 1 H).

3. Discussion

4-Methoxy-3-penten-2-one has been prepared by Awang using methanol and sulfuric acid.[2] He also determined the stereochemistry by NMR solvent shift data and observation of nuclear Overhauser effects. Our preparation is a convenient, one pot procedure. The title compound is useful for effecting the overall γ-alkylation of enones[3] and has been used in a synthesis of prostaglandins.[4]

1. Department of Chemistry, Iowa State University, Ames, IA 50011.
2. Awang, D. V. C. *Can. J. Chem.* **1971**, *49*, 2672.
3. Stork, G.; Kraus, G. A. *J. Am. Chem. Soc.* **1976**, *98*, 2351.
4. Stork, G.; Kraus, G. A. *J. Am. Chem. Soc.* **1976**, *98*, 6747.

Appendix
Chemical Abstracts Nomenclature (Collective Index Number); (Registry Number)

4-Methoxy-3-penten-2-one: 3-Penten-2-one, 4-methoxy- (8,9); (2845-83-2)

2,4-Pentanedione (8,9); (123-54-6)

3-HYDROXY-1-CYCLOHEXENE-1-CARBOXALDEHYDE

(1-Cyclohexene-1-carboxaldehyde, 3-hydroxy-)

Submitted by H. L. Rigby, M. Neveu, D. Pauley, B. C. Ranu, and T. Hudlicky.[1]
Checked by Denis R. St. Laurent and Leo A. Paquette.

1. Procedure

A. *1-(1,3-Dithian-2-yl)-2-cyclohexen-1-ol* (2). To a suspension of 1,3-dithiane (Note 1) (12 g, 0.1 mol) in dry tetrahydrofuran (100 mL) (Note 2) at -78°C is added butyllithium (40 mL, 2.5 M, 0.1 mol). The reaction is stirred for 2 hr. Initially the dithiane dissolves, followed by precipitation of the lithio salt. After 2 hr, 2-cyclohexen-1-one (9.6 g, 0.1 mol) (Note 3) in dry tetrahydrofuran (20 mL) is added dropwise. After about half of the cyclohexenone is added, the mixture becomes homogeneous. After the addition is complete, the reaction is stirred for an additional 30 min at -78°C and then stored for 18 hr at 0°C. The solution is concentrated to one-fourth volume under reduced pressure. Water (100 mL) is added and the mixture is extracted with ether (3 x 50 mL). The extract is dried over sodium sulfate and evaporated to give an oil which is vacuum distilled to give 14.0-14.7 g (65-68%) of the protected aldehyde 2, bp 149-153°C (0.8-1.0 mm) (Note 4).

B. *3-Hydroxy-1-cyclohexene-1-carboxaldehyde* (**3**). The hydroxy thioacetal **2** (5.5 g, 0.025 mol) in 25 mL of tetrahydrofuran is added dropwise to a mechanically-stirred suspension of red mercuric oxide (11 g, 0.051 mol) (Note 5) and boron trifluoride etherate (7.2 g, 0.051 mol) (Note 6) in refluxing 15% aqueous tetrahydrofuran (50 mL). After the addition is complete, the mixture is stirred (Note 7) at reflux for an additional 2 hr. Another 5.5 g of red mercuric oxide is added and the reaction is stirred at reflux for 1 hr. The reaction is cooled to room temperature and ether (150 mL) is added followed by 50 mL of brine. The mixture is filtered and the residue washed with ether (3 x 50 mL). The organic layer is separated and washed with saturated sodium bicarbonate solution (2 x 50 mL) and brine (1 x 50 mL). The organic layer is dried over sodium sulfate and the solvent evaporated to leave a residual oil. This oil is purified by medium pressure liquid chromatography (silica gel, elution with 40% ethyl acetate in hexane) to give 1.5-1.6 g (47-50%) of the aldehyde 3, bp 125°C (0.075 mm) (Kugelrohr) (Note 8).

2. Notes

1. Dithiane was obtained from Aldrich Chemical Company, Inc. and was used without purification.

2. Tetrahydrofuran was distilled under nitrogen from potassium and benzophenone.

3. 2-Cyclohexen-1-one was obtained from Aldrich Chemical Company, Inc. and used as received.

4. The submitters report bp 157°C (1 mm). The spectral properties of 2 are as follows: IR (neat) cm^{-1}: 3450 (br), 3030, 2940, 2900 (sh), 2830, 1645 (w), 1425, 1280, 1185, 1085, 985; ^{1}H NMR (CDCl$_3$, 300 MHz) δ: 1.57-2.08 (m, 8 H), 2.40 (s, 1 H), 2.74-2.90 (m, 4 H), 4.17 (s, 1 H), 5.68 (d, 1 H, J = 10.05), 5.85-5.91 (m, 1 H); ^{13}C NMR (CDCl$_3$, 20 MHz) δ: 18.53, 24.98, 25.75, 30.35, 30.54, 33.03, 59.54, 71.75, 129.43, 132.18; MS: M$^+$ 198, 120, 119 (base), 97, 91, (no M$^+$ peak).

5. Mercuric oxide (red) was purchased from Aldrich Chemical Company, Inc. and used without purification.

6. Boron trifluoride etherate was distilled from calcium hydride at aspirator pressure.

7. Mechanical stirring was found to be essential and in some cases addition of sea sand to the reaction mixture reduced clogging of the reagents and led to higher yields.

8. The submitters report bp 70°C (10^{-4} mm). The spectral properties of **3** are as follows: IR (neat) cm^{-1}: 3400 (br), 2950, 2870, 2720, 1675, 1435, 1305, 1180, 1135, 1070, 1045, 995, 965, 925; ^{1}H NMR (CDCl$_3$, 300 MHz) δ: 1.47-1.54 (m, 2 H), 1.71-1.79 (m, 1 H), 1.90-1.98 (m, 1 H), 2.06 (s, 2 H), 3.49 (br s, 1 H), 4.37 (br s, 1 H), 6.63 (s 1 H), 9.34 (s, 1 H); ^{13}C NMR (CDCl$_3$, 75 MHz) δ: 18.54, 20.94, 31.35, 65.84, 77.43, 141.63, 150.88, 194.49; MS: M$^+$ 97, (base), 79, 69, 55, 41.

3. Discussion

Preparation of a compound such as 3 is useful where extensive functionalization of a cyclic enone is required. In addition to 3, 4,4-dimethyl-3-hydroxy-1-cyclopentene-1-carboxaldehyde was prepared using the same procedure from 4,4-dimethyl-2-cyclopenten-1-one.[2] The Table below gives yields and physical properties for this compound which has been used as a starting material in the syntheses of coriolin and pentalenic acid.[3]

TABLE

PREPARATION OF 4,4-DIMETHYL-3-HYDROXY-1-CYCLOPENTENE-1-CARBOXALDEHYDE

Enone	Dithiane	Hydroxy Aldehyde
	79% bp 130-132°C/0.05 mm	55%

1. Department of Chemistry, Virginia Polytechnic Institute and State University, Blacksburg, VA 24061.
2. Magnus, P. D.; Nobbs, M. S. *Synth. Commun.* **1980**, *10*, 273.
3. Hudlicky, T.; Kwart, L. D.; Tiedje, M. H.; Ranu, B. C.; Short, R. P.; Frazier, J. O.; Rigby, H. L. *Synthesis* **1986**, 716.

Appendix

Chemical Abstracts Nomenclature (Collective Index Number);
(Registry Number)

3-Hydroxy-1-cyclohexene-1-carboxaldehyde: 1-Cyclohexene-1-carboxaldehyde, 3-hydroxy- (10); (67252-14-6)

1-(1,3-Dithian-2-yl)-2-cyclohexen-1-ol: 2-Cyclohexen-1-ol, 1-(1,3-dithian-2-yl)- (9); (53178-46-4)

Cyclohexenone: 2-Cyclohexen-1-one (8,9); (930-68-7)

4,4-Dimethyl-2-cyclopenten-1-one: 2-Cyclopenten-1-one, 4,4-dimethyl- (8,9); (22748-16-9)

SYNTHESIS OF CYCLOBUTANONES VIA 1-BROMO-1-ETHOXYCYCLOPROPANE:
(E)-2-(1-PROPENYL)CYCLOBUTANONE
(Cyclobutanone, 2-(1-propenyl)-, (E))

A.

B.

C.

Submitted by Scott A. Miller and Robert C. Gadwood.[1]
Checked by Jeffrey A. McKinney and Leo A. Paquette.

1. Procedure

A. *1-Bromo-1-ethoxycyclopropane*.[2] A 500-mL, round-bottomed flask equipped with a magnetic stirring bar and a calcium sulfate drying tube is charged with 84.1 g (0.483 mol) of 1-ethoxy-1-trimethylsiloxycyclopropane.[3] Phosphorus tribromide (35.6 mL, 103 g, 0.379 mol) (Note 1) is added at room temperature with brisk stirring, followed by a catalytic amount (0.5 ml) of 48% aqueous hydrobromic acid (Note 2). The resulting clear, pale yellow solution is stirred for 6 hr (Note 3). After the stirring bar is removed, the reaction mixure is distilled by Kugelrohr apparatus at aspirator vacuum (10 mm) from 25°C to 70°C to afford crude 1-bromo-1-ethoxycyclopropane (Notes 4 and 5). The crude product is dissolved in 300 mL of pentane in a 1-L Erlenmeyer flask and the resulting solution is chilled to -20°C in a dry ice-ethanol: water (30:70) bath. While the temperature of the solution is maintained below 25°C, 300 mL of saturated, aqueous sodium carbonate is carefully added (Note 6). The layers are carefully shaken and separated, and the aqueous phase is extracted with 100-mL of pentane. The organic layer is dried over magnesium sulfate, filtered, and most of the pentane is removed by distillation through a 15-cm Vigreux column at atmospheric pressure. The residue is transferred to a smaller distillation flask and distilled through the same column under aspirator vacuum to afford 47.0-57.6 g (59-72%) of 1-bromo-1-ethoxycyclopropane as a colorless liquid (bp 35-43°C, 10 mm) (Notes 7 and 8).

Caution! Because of the relatively large amount of pyrophoric tert-butyllithium involved, the following preparation should be performed in a hood behind a safety shield.

B. *(E)-1-Ethoxy-1-(1-hydroxy-2-butenyl)cyclopropane.* A 1-L, three-necked flask is equipped with a gas inlet adapter, a septum, a 250-mL graduated addition funnel capped with a septum, and a magnetic stirring bar (Note 9). The flask is charged with 500 mL of anhydrous diethyl ether (Note 10) and cooled to -78°C under nitrogen atmosphere. The addition funnel is charged with 177 mL (19.2 g, 0.30 mol) of tert-butyllithium (Note 11), transferred from the reagent bottle via a stainless steel cannula under positive nitrogen pressure. The tert-butyllithium is added dropwise to the stirred diethyl ether over approximately 20 min while the cooling bath is maintained at -78°C. After the addition is complete, 26.4 g, (0.16 mol) of freshly prepared 1-bromo-1-ethoxycyclopropane is added to the reaction over about 5 min by syringe. The resulting cloudy, colorless or light yellow reaction mixture is stirred for 20-25 min, and a solution of 7.0 g (0.10 mol) of crotonaldehyde (Note 12) in 50 mL of anhydrous diethyl ether (chilled to -78°C) is added via a stainless steel cannula under positive nitrogen pressure. The reaction mixture is stirred at -78°C for an additional 10 min, warmed to 0°C in an ice bath, and carefully quenched with 100 mL of saturated, aqueous ammonium chloride. The layers are shaken and separated and the aqueous phase is extracted with 100 mL of diethyl ether. The combined organic layers are dried over magnesium sulfate and filtered. After the crude adduct is concentrated on a rotary evaporator, it is filtered through a 10-cm pad of silica gel (Note 13) in a sintered-glass funnel with 10% ethyl acetate in hexane. After concentrating again on a rotary evaporator, the crude adduct is obtained as a pale yellow oil (14.2-15.6 g) (Note 14). This material is not further purified, but is used directly in the next reaction.

C. *(E)-2-(1-Propenyl)cyclobutanone.* To a 1-L, round-bottomed flask equipped with a magnetic stirring bar is added 15.3 g (0.098 mol) of (E)-1-ethoxy-1-(1-hydroxy-2-butenyl)cyclopropane, 500 mL of reagent grade diethyl ether and 6.6 mL (4.3 g, 0.049 mol) of 48% aqueous fluoboric acid (Note 15). After the reaction mixture is stirred for 15 min at room temperature, it is quenched with 60 mL (0.06 mol) of 1 M aqueous sodium carbonate. The layers are carefully shaken and separated, and the organic phase is washed with three 125-mL portions of water (Note 16). The combined aqueous layers are extracted with 100 mL of diethyl ether and the organic phase is dried over magnesium sulfate and filtered. The filtrate is concentrated on a rotary evaporator without external heating and the residue is distilled through a 10-cm Vigreux column under aspirator vacuum. The product, 7.2-8.3 g (66-75% yield from crotonaldehyde), is obtained as a colorless oil, bp 61-65°C (10 mm) (Note 17).

2. Notes

1. Phosphorus tribromide was obtained from the Aldrich Chemical Company, Inc. and used without further purification.

2. The addition of a catalytic amount of hydrobromic acid was frequently found to be necessary to initiate the reaction, especially if the phosphorus tribromide is of high purity. Upon addition of the hydrobromic acid, the reaction warms noticeably.

3. The course of the reaction is most conveniently followed by ^{1}H NMR analysis of a drop of the reaction mixture in carbon tetrachloride. The downfield quartet of the starting ketal (3.52 ppm) is replaced by a clean quartet at 3.62 ppm from the product. The checkers have found by this technique that reaction is complete in much less than 6 hr.

4. The crude product thus obtained also contains bromotrimethylsilane and hydrobromic acid.

5. *Caution!* After distillation the Kugelrohr apparatus should first be cooled and then carefully vented to an atmosphere of nitrogen since traces of elemental phosphorus may be present in the pot residue and may ignite if exposed to air while still hot.

6. This step neutralizes the hydrobromic acid and bromotrimethylsilane present in the product. Therefore, addition of the aqueous sodium carbonate solution is exothermic and causes vigorous carbon dioxide evolution. Cooling at this stage helps prevent hydrolysis of the product.

7. A low boiling, silicon-containing fraction is also collected below 37°C (28 mm). The presence of a singlet at 0.10 ppm in the ^{1}H NMR of the product indicates contamination by this low boiling fraction. Small amounts of this impurity do not seem to interefere in subsequent reactions of the 1-bromo-1-ethoxycyclopropane.

8. 1-Bromo-1-ethoxycyclopropane is relatively unstable at room temperature, but can be stored for several months at -20°C with only slight decomposition. Spectral data for 1-bromo-1-ethoxycyclopropane are as follows: IR (neat) cm^{-1}: 3100 (w), 2985 (s), 2935 (m), 2885 (m), 1445 (m), 1300 (s), 1160 (s), 1060 (s), 795 (s); ^{1}H NMR (CCl$_4$) δ: 1.17 (m, 7 H), 3.53 (q, 2 H, J = 8); MS (15 eV), m/e 164/166 (M$^+$), 136/138 (base), 85, 57.

9. The glassware was dried in an oven overnight at 110°C and assembled while hot under nitrogen flow.

10. Diethyl ether was dried by distillation from sodium metal/benzophenone.

11. *Caution! tert-Butyllithium is extremely pyrophoric and should only be handled on a large scale by experienced personnel.* tert-Butyllithium was obtained from the Aldrich Chemical Company, Inc. as a 1.7 M solution in pentane. In general, this material was used as received without titration.

12. Crotonaldehyde was obtained from The Matheson Company, Inc., and is also available (99+% grade) from the Aldrich Chemical Company, Inc.

13. Merck Silica Gel 60 (230-400 mesh) was obtained from the Aldrich Chemical Company, Inc. Filtration through silica gel removes residual inorganic salts (mostly lithium chloride) which may interfere in the subsequent rearrangement step.

14. Spectral data for (E)-1-ethoxy-1-(1-hydroxy-2-butenyl)cyclopropane are as follows: ^{1}H NMR (CDCl$_3$) δ: 0.68 (m, 4 H), 1.12 (t, 3 H, J = 6), 1.64 (d, 3 H, J = 5), 2.46 (s, 1 H), 3.54 (m, 2 H), 4.15 (d, 1 H, J = 6), 5.52 (m, 2 H). Occasionally, a minor impurity is formed as a result of the addition of tert-butyllithium to crotonaldehyde (singlet at 0.89 ppm in the ^{1}H NMR). This side reaction occurs because of the presence of unreacted tert-butyllithium and is best avoided by using the indicated ratio of tert-butyllithium to 1-bromo-1-ethoxycyclopropane. The checkers were unable to remove this impurity by fractional distillation.

15. Fluoboric acid was obtained as a 48 wt % aqueous solution from the Aldrich Chemical Company, Inc. Based upon its density, this solution was calculated to be approximately 7.4 M in HBF$_4$. The checkers used 5.4 mL (0.049 mol) of 60% fluoboric acid.

16. Washing with water helps to remove the ethanol generated in the course of the rearrangement. For higher boiling cyclobutanones, where the ethanol can easily be removed during distillation, this step is unnecessary.

17. Spectral data for (E)-2-(1-propenyl)cyclobutanone are as follows: IR (CCl$_4$) cm^{-1}: 2960 (s), 1780 (s), 1660 (w), 1450 (m); ^{1}H NMR (CCl$_4$) δ: 1.62 (m, 3 H), 2.17 (m, 2 H), 2.88 (m, 2 H), 3.73 (m, 1 H), 5.37 (m, 2 H). The product was contaminated by an alcoholic impurity to the extent of 6-11%.

3. Discussion

Cyclobutanones have attained a position of considerable synthetic importance in recent years. In addition to being important synthetic targets themselves, they serve as useful precursors of five-,[4] six-,[5] and eight-membered[6] rings, as well as of a variety of highly functionalized acyclic fragments.[7,8]

In general, cyclobutanones are synthesized by either ketene cycloadditions or by ring expansions of cyclopropyl precursors. For the synthesis of simple α-substituted monocyclic cyclobutanones, the latter method is usually employed, and a variety of approaches have been used to prepare the required cyclopropyl intermediates.

Vinylcyclopropanols have been prepared by the addition of alkenyl Grignard reagents to a variety of cyclopropanone equivalents.[9] Upon treatment with acid, the vinylcyclopropanols rearrange to α-substituted cyclobutanones. Alternatively, a variety of α-heteroatom-substituted cyclopropyllithium reagents have been developed. These react with aldehydes and ketones to afford cyclopropylcarbinols which also rearrange to cyclobutanones under acid catalysis.[8,10,11] Lastly, vinylcyclopropanols and cyclopropylcarbinols have been prepared by the cyclopropanation of enol silyl ethers and allylic alcohols.[12]

There are several advantages to the procedure described here for the synthesis of α-substituted cyclobutanones. The preparation of 1-bromo-1-ethoxycyclopropane is convenient and can be accomplished in good overall yield in only two steps from commercially available ethyl 3-chloropropionate. Metalation of 1-bromo-1-ethoxycyclopropane is rapid and reproducible on a large scale and (1-ethoxy)cyclopropyllithium adds cleanly to a wide variety of ketones and aldehydes. Finally, rearrangement of the cyclopropylcarbinol adducts occurs smoothly and in high yield.

The preparation of 1-bromo-1-ethoxycyclopropane is based on a literature report of the synthesis of 1-bromo-1-methoxycyclopropane from 1-methoxy-1-trimethylsiloxycyclopropane using phosphorus tribromide in pyridine.[13] In our hands, reaction of 1-ethoxy-1-trimethylsiloxycyclopropane under these conditions afforded none of the bromide.

The title cyclobutanone has been prepared previously by the addition of (1-phenylthio)cyclopropyllithium to crotonaldehyde followed by rearrangement with anhydrous stannic chloride in methylene chloride.[11] However, in our experience, the procedure described here is much more convenient and reproducible on a large scale.

As shown in the Table, a wide variety of α-substituted cyclobutanones have been prepared by the general method described here.[14] The time required for rearrangement of the intermediate cyclopropylcarbinols varies from less than 5 min for entry 2 to 48 hr for entry 10. With most enones and enals, only 1,2-addition is observed, but in two cases (entries 3 and 4), a significant amount of the 1,4-adduct was also produced. The increased 1,4-addition seen in entry 3 apparently occurs because of steric factors, whereas that seen in entry 4 presumably occurs because of chelation of the organolithium to the benzyl ether oxygen.

1. Department of Chemistry, Northwestern University, Evanston, IL 60201.
2. Gadwood, R. C. *Tetrahedron Lett.* **1984**, *25*, 5851; Gadwood, R. C.; Rubino, M. R.; Nagarajan, S. C.; Michel, S. T. *J. Org. Chem.* **1985**, *50*, 3255.
3. Salaün, J.; Marguerite, J. *Org. Synth.* **1985**, *63*, 147.
4. Gadwood, R. C. J. Org. Chem. **1983**, *48*, 2098, and references therein.

5. Wilson, S. R.; Mao, D. T. *J. Chem. Soc., Chem. Commun.* **1978**, 479; Danheiser, R. L.; Martinez-Davila, C.; Sard, H. *Tetrahedron* **1981**, *37*, 3943; Cohen, T.; Bhupathy, M.; Matz, J. R. *J. Am. Chem. Soc.* **1983**, *105*, 520.

6. Gadwood, R. C.; Lett, R. M. *J. Org. Chem.* **1982**, *47*, 2268; Paquette, L. A.; Andrews, D. R.; Springer, J. P. *J. Org. Chem.* **1983**, *48*, 1147.

7. Trost, B. M. *Acc. Chem. Res.* **1974**, *7*, 85; Trost, B. M.; Bogdanowicz, M. J.; Kern, J. *J. Am. Chem. Soc.* **1975**, *97*, 2218; Trost, B. M.; Preckel, M.; Leichter, L. M. *J. Am. Chem. Soc.* **1975**, *97*, 2224.

8. For selected references to other uses of cyclobutanones in synthesis, see: Trost, B. M.; Ornstein, P. L. *J. Org. Chem.* **1983**, *48*, 1131.

9. Ollivier, J.; Salaün, J. *Tetrahedron Lett.* **1984**, *25*, 1269; Salaün, J. *Chem. Rev.* **1983**, *83*, 619; Salaün, J.; Garnier, B.; Conia, J. M. *Tetrahedron* **1974**, *30*, 1413; Salaün, J.; Conia, J. M. *Tetrahedron Lett.* **1972**, 2849; Wasserman, H. H.; Hearn, M. J.; Cochoy, R. E. *J. Org. Chem.* **1980**, *45*, 2874; Wasserman, H. H.; Hearn, M. J.; Haveaux, B.; Thyes, M. *J. Org. Chem.* **1976**, *41*, 153; Wasserman, H. H.; Cochoy, R. E.; Baird, M. S. *J. Am. Chem. Soc.* **1969**, *91*, 2375; Wasserman, H. H.; Clagett, D. C. *J. Am. Chem. Soc.* **1966**, *88*, 5368.

10. Dammann, R.; Seebach, D. *Chem. Ber.* **1979**, *112*, 2167; Braun, M.; Dammann, R.; Seebach, D. *Chem. Ber.* **1975**, *108*, 2368; Braun, M.; Seebach D. *Angew. Chem., Intern. Ed. Engl.* **1974**, *13*, 277; Hiyama, T.; Takehara, S.; Kitatani, K.; Nozaki, H. *Tetrahedron Lett.* **1974**, 3295; Halazy, S.; Zutterman, F.; Krief, A. *Tetrahedron Lett.* **1982**, *23*, 4385; Trost, B. M.; Bogdanowicz, M. J. *J. Am. Chem. Soc.* **1973**, *95*, 5311; Johnson, C. R.; Katekar, G. F.; Huxol, R. F.; Janiga, E. R. *J. Am. Chem. Soc.* **1971**, *93*, 3771; Johnson, C. R.; Janiga, E. R. *J. Am. Chem. Soc.* **1973**, *95*, 7692;

Cohen, T.; Sherbine, J. P.; Matz, J. R.; Hutchins, R. R.; McHenry, B. M.; Willey, P. R. *J. Am. Chem. Soc.* **1984**, *106*, 3245; Cohen, T.; Matz, J. R. *J. Am. Chem. Soc.* **1980**, *102*, 6900.

11. Trost, B. M.; Keeley, D. E.; Arndt, H. C.; Rigby, J. H.; Bogdanowicz, M. J. *J. Am. Chem. Soc.* **1977**, *99*, 3088.

12. Girard, C.; Amice, P.; Barnier, J. P.; Conia, J. M. *Tetrahedron Lett.* **1974**, 3329; Wenkert, E.; Arrhenius, T. S. *J. Am. Chem. Soc.* **1983**, *105*, 2030 and references cited therein.

13. van Tilborg, M. W. E. M.; van Doorn, R.; Nibbering, N. M. M. *J. Am. Chem. Soc.* **1979**, *101*, 7617.

14. Entries 2, 9, and 10 of Table I have been previously reported.[2] Entry 1 was carried out by Amy J. DeWinter. Entry 3 was carried out by Scott A. Miller. Entry 4 was carried out by Mark R. Rubino. Entries 5-8 were carried out by Ishwar M. Mallick.

TABLE

CYCLOBUTANONE SYNTHESIS VIA 1-BROMO-1-ETHOXYCYCLOPROPANE

Entry	Ketone/aldehyde	Cyclobutanone	Yield (%)
1	2-cyclohexenone	spiro cyclobutanone with cyclohexene	97
2	mesityl oxide	2-methyl-2-(2-methylpropenyl)cyclobutanone	86
3	ethyl vinyl ketone	2-ethyl-2-vinylcyclobutanone	40
4	(E)-5-benzyloxy-3-penten-2-one	2-methyl-2-(3-benzyloxy-1-propenyl)cyclobutanone	30
5	2-methylenehexanal	2-(1-methylenepentyl)cyclobutanone	65
6	benzaldehyde	2-phenylcyclobutanone	71
7	cinnamaldehyde	2-styrylcyclobutanone	79
8	(E)-4-(TBDMSoxy)-2-nonenal	2-[(E)-4-(TBDMSoxy)-2-nonenyl]cyclobutanone	74
9	cyclohexanone	spiro[3.5]nonan-1-one	81
10	heptanal	2-hexylcyclobutanone	81

220

Appendix

Chemical Abstracts Nomenclature (Collective Index Number);

(Registry Number)

1-Bromo-1-ethoxycyclopropane: Cyclopropane, 1-bromo-1-ethoxy- (11); (95631-62-2)

(E)-2-(1-Propenyl)cyclobutanone: Cyclobutanone, 2-(1-propenyl)-, (E)- (10); (63049-06-9)

1-Ethoxy-1-trimethylsiloxycyclopropane: Silane, [(1-ethoxycyclopropyl)oxy]-trimethyl- (8,9); (27374-25-0)

4-CHLORINATION OF ELECTRON-RICH BENZENOID COMPOUNDS:
2,4-DICHLOROMETHOXYBENZENE
(Benzene, 2,4-dichloro-1-methoxy-)

A. morpholine $\xrightarrow{\text{NaOCl} / H_2O}$ N-chloromorpholine

B. 2-chloroanisole + N-chloromorpholine $\xrightarrow{80\% H_2SO_4}$ 2,4-dichloroanisole + morpholine

Submitted by John R. Lindsay Smith, Linda C. McKeer, and Jonathan M. Taylor.[1]
Checked by Yasushi Morita and Ryoji Noyori.

1. Procedure

A. N-Chloromorpholine. A 500-mL, three-necked, round-bottomed flask equipped with a dropping funnel, mechanical stirrer and thermometer is charged with 250 mL of 1.5 M sodium hypochlorite solution (Note 1). The solution is stirred and the temperature is maintained below 10°C while 30 mL (0.34 mol) of morpholine (Note 2) is added dropwise. The resulting mixture is stirred for 5 min before the N-chloromorpholine is extracted with four 50-mL portions of diethyl ether. The combined ether extracts are dried over anhydrous magnesium sulfate and concentrated with a rotary evaporator (Note 3). The concentrate is distilled at reduced pressure (Note 4) to afford 35.5-36.5 g (86-88%) of N-chloromorpholine, bp 63-64°C (36-38 mm) (Note 5).

B. *2,4-Dichloromethoxybenzene.* A 500-mL, three-necked, round-bottomed flask equipped with a dropping funnel, mechanical stirrer and thermometer is charged with 250 mL of 80% (v/v) sulfuric acid (Note 6) and cooled in an ice bath before 16 g (0.11 mol) of 2-chloromethoxybenzene (Note 7) is added with stirring. The stirring and cooling are maintained while 14.5 g (0.12 mol) of N-chloromorpholine is added dropwise (Note 8). The cooling bath is removed and stirring is continued for 1 hr. The reaction mixture is carefully poured into a mixture of 150 mL of distilled water and 100 g of crushed ice in a 1-L flask cooled at 0°C (Note 9). The aromatic products are extracted with a 100-mL portion, followed by four 50-mL portions, of diethyl ether. The combined ether extracts are washed with 100 mL of water containing 0.5 g of potassium iodide, 2 g of sodium thiosulfate and 2 mL of acetic acid (Note 10) followed by 50 mL of 8% (w/v) aqueous sodium hydroxide (Note 11), dried over anhydrous magnesium sulfate, and concentrated with a rotary evaporator. The concentrate is distilled under reduced pressure to afford 15.2-16.0 g (77-81%) of 2,4-dichloromethoxybenzene, bp 110-111°C (10 mm) [lit.[2] bp 125°C (10 mm), 233°C (740 mm)]. The product after distillation is 98.9-99.2% pure, the major impurities being 2,6-dichloromethoxybenzene (0.4-0.5%) and 2,4,6-trichloromethoxybenzene (0.4-0.6%) (Notes 12 and 13). If the above purity is insufficient, it can be improved to >99.9% by recrystallization (Notes 14 and 15).

2. Notes

1. Solutions of sodium hypochlorite of different concentration can be used with a corresponding change in the volume. The submitters purchased sodium hypochlorite solution from BDH Chemicals Ltd., England. The material

initially contains 10-14% available chlorine, but it deteriorates on standing over a period of weeks. The checkers used the material (chlorine content 9-14%) purchased from Nakarai Chemicals, Ltd., Japan.

2. Gold Label grade morpholine (99+%) was obtained from Aldrich Chemical Company, Inc., and used as supplied.

3. Since N-chloromorpholine has a low boiling-point, the water bath temperature should not exceed 30°C.

4. It is recommended that a water or oil bath, or a hot air blower, be used for this distillation to avoid the risk of local overheating.

5. N-Chloromorpholine should be handled with extreme care at all times. On standing at room temperature it slowly decomposes, forming crystals of morpholine hydrochloride. However, it can be stored for several weeks at -18°C. Vigorous decomposition of N-chloromorpholine has been reported when it is distilled at atmospheric pressure.[3] The checkers removed the small quantity of salt contamination by filtration through a glass filter and used the pure liquid in the subsequent chlorination reaction.

6. Trifluoroacetic acid, 100 mL, obtainable from Aldrich Chemical Company, Inc., can be used instead of aqueous sulfuric acid (see Discussion, part 3).

7. 2-Chloromethoxybenzene was obtained from Aldrich Chemical Company, Inc., and was distilled prior to use, bp 195-196°C or 112°C (41 mm).

8. Since the dissolution of N-chloromorpholine in sulfuric acid (or trifluoroacetic acid) and the subsequent reaction between protonated N-chloromorpholine and 2-chloromethoxybenzene are both exothermic processes, the addition of the chloroamine should be carried out at such a rate as to keep the reaction temperature below 5°C. The checkers found that a reaction run at 8°C gave product of only 93% purity.

9. If the reaction is carried out in trifluoroacetic acid, the product mixture is made basic by adding it cautiously, with cooling and stirring, to a cold solution of 50 g of sodium hydroxide in 150 mL of distilled water. The aromatic products are then extracted with diethyl ether as described in the main text.

10. If trifluoroacetic acid is used, more acetic acid may be required to ensure that the aqueous layer is acidic. Should any iodine remain, more sodium thiosulfate should be added until all of the iodine has been converted to iodide.

11. The aqueous layer should remain basic after washing with sodium hydroxide. If it is still acidic, this wash should be repeated.

12. If trifluoroacetic acid is used as solvent, the purity is 98-99% and the impurities are mainly 2,6-dichloro- and 2,4,6-trichloromethoxybenzene.

13. The purity of the product can be determined by gas-liquid chromatography using a column packed with 10% (w/w) Carbowax 20 M on Celite (80-100 mesh) at 195°C, nitrogen carrier gas flow rate 35 mL min^{-1}.

14. Eighteen grams of 2,4-dichloromethoxybenzene is dissolved in 20 mL of light petroleum ether and chilled to -18°C. Crystallization can be induced by either scratching or seeding. The mixture is kept at -18°C for 1 hr to maximize the yield before the crystals are filtered with a Buchner funnel and washed with 10 mL of chilled light petroleum ether. The crystals are sucked dry, and then dried in a vacuum desiccator. The recrystallized yield of 2,4-dichloromethoxybenzene is 12.8 g (55-58% overall), mp 25.5-26.5°C, lit.[2] mp 28°C.

15. The product had the following spectral properties. IR (neat) cm^{-1}: 1483, 1288, 1254, 1055, 700; ^{1}H NMR (CCl$_4$, 60 MHz) δ: 3.77 (s, 3 H, OCH$_3$), 6.70 (d, 1 H, J = 9.0, H$_6$), 7.05 (dd, 1 H, J = 9.0 and 2.5, H$_5$), 7.27 (d, 1 H,

J = 2.5, H$_3$); ^{13}C NMR (CDCl$_3$, 22.5 MHz) δ: 56.2 (q), 112.7 (d), 123.3 (s), 125.6 (s), 127.5 (d), 129.9 (d), 153.9 (s); mass spectrum (70 eV) m/e (rel. intensity): 178 (M$^+$ + 2, 66), 176 (M$^+$, 100), 163 (47), 161 (58), 135 (23), 133 (43); mass spectrum calcd for C$_7$H$_6$OCl$_2$ (M$^+$): 175.9797, found: 175.9809.

3. Discussion

Chlorine or hypochlorous acid has been used traditionally for the chlorination of aromatic compounds and, when required, the reactivity of these reagents can be increased with a Lewis or protic acid, respectively.[4] However, these reactions are rarely selective for one monochlorinated product (site-selective[5]) and, furthermore, with some substrates di- and poly-chlorination can also occur. The increasing need for isomerically pure chloroaromatics in recent years has led to the development of more selective chlorinating agents, particularly for electron-rich aromatic compounds (e.g., phenol).[6] In this respect the submitters have found that N-chloro-dialkylamines in strongly acidic solution are efficient and very selective mono-chlorinating agents for aromatic compounds containing a π-donor (+M) substituent.[7] Thus, normally the addition of the N-chloroamine to an equimolar quantity of the substrate in acid leads rapidly and almost exclusively to the para-chlorinated product (Table I). Although, for reasons of economy most of the reactions have been studied on a small scale (<1 g of substrate), the submitters have had no difficulty in scaling up the chlorinations to use 20 g of substrate. The two acidic media that have been used with success are trifluoroacetic acid and aqueous sulfuric acid [commonly 80% (v/v) sulfuric acid]. The advantages of the former are that the reactions are homogeneous, can if necessary be carried out at low temperature (below

0°C) and can be monitored readily by ^{1}H NMR spectroscopy. However, trifluoroacetic acid is relatively expensive and is highly toxic (the reactions must be carried out in a well-ventilated hood). In situations where these disadvantages outweigh the advantages, aqueous sulfuric acid is generally a cheap and less toxic alternative. The fact that the reactions in aqueous sulfuric acid are not homogeneous is not a serious problem. Thus, with efficient stirring the chlorinations occur rapidly; furthermore, solid substrates can be added as solutions in diethyl ether (e.g., with N-chloromorpholine, phenol gave 93% of 4-chloro- and 7% of 2-chlorophenol, and 2-methylphenol gave 95% of 4-chloro- and 5% of 6-chloro-2-methylphenol). The major disadvantage in the use of aqueous sulfuric acid arises with the most reactive substrates (e.g., some phenols) from competing aromatic sulfonation. However, this can be reduced to a minor side-reaction by keeping the reaction mixture cold (below 8°C the 80% sulfuric acid reaction mixtures will begin to freeze) and by minimizing the time between the addition of the substrate and of the chloroamine to the aqueous sulfuric acid.

The structure of the N-chlorodialkylamine markedly affects its reactivity and to a lesser extent its selectivity (Table II). Thus with 2-chloromethoxybenzene as substrate, N-chloromorpholine is approximately 17,000 times more reactive than N-chloropiperidine and yet it is only slightly less selective for para-chlorination of methoxybenzene. For most substrates the shorter reaction times (less chance of other side reactions) of the more reactive N-chloroamines more than compensates for any small decrease in selectivity.

1. Department of Chemistry, University of York, York YO1 5DD, England.
2. "Dictionary of Organic Compounds", 5th ed.; Buckingham, J., Ed.; Chapman and Hall: New York, 1982; Vol. 2, p. 1767.
3. Henry, R. A.; Dehn, W. M. *J. Am. Chem. Soc.* **1950**, *72*, 2280.
4. de la Mare, P. B. D. "Electrophilic Halogenation", Cambridge Univ. Press: Cambridge, England, 1976; March, J. "Advanced Organic Chemistry", 2nd ed.; McGraw Hill: New York, 1977; Chapter 11.
5. Fleming, I. "Frontier Orbitals and Organic Chemical Reactions", J. Wiley & Sons: New York, 1976; p. 165.
6. Crocker, H. P.; Walser, R. *J. Chem. Soc. C* **1970**, 1982; Guy, A.; Lemaire, M.; Guetté, J.-P. *Tetrahedron* **1982**, *38*, 2339 and 2347; Onyiriuka, S. O.; Suckling, C. J. *J. Chem. Soc., Chem. Commun.* **1982**, 833; Watson, W. D. *J. Org. Chem.* **1985**, *50*, 2145.
7. Lindsay Smith, J. R.; McKeer, L. C. *Tetrahedron Lett.* **1983**, *24*, 3117.
8. Fuller, S. E.; Lindsay Smith, J. R.; Norman, R. O. C.; Higgins, R. *J. Chem. Soc., Perkin Trans. II*, **1981**, 545.

Appendix

Chemical Abstracts Nomenclature (Collective Index Number);

(Registry Number)

2,4-Dichloromethoxybenzene: Anisole, 2,4-dichloro- (8); Benzene, 2,4-dichloro-1-methoxy- (9); (553-82-2)

N-Chloromorpholine: Morpholine, 4-chloro- (8,9); (23328-69-0)

Morpholine (8,9); (110-91-8)

2-Chloromethoxybenzene: Anisole, o-chloro- (8); Benzene, 1-chloro-2-methoxy- (9); (766-51-8)

TABLE I

YIELD AND PRODUCT DISTRIBUTIONS FROM THE CHLORINATION
OF AROMATIC COMPOUNDS IN TRIFLUOROACETIC ACID[a]

Substrate	Chlorinating Agent	Yield[b] (%)	Product	Product Distribution (%)
C_6H_5OMe	NCP[c]	97	2-chloromethoxybenzene	1
			4-chloromethoxybenzene	99
C_6H_5OH	NCP	98	2-chlorophenol	3
			4-chlorophenol	97
2-methylphenol (OH, CH₃)	NCP	84	6-chloro-2-methylphenol	1.5
			4-chloro-2-methylphenol	98.5
1,2-dimethoxybenzene (OCH₃, OCH₃)	NCTA[d]	80	4-chloro-1,2-dimethoxy-benzene	100
3-methylphenol (OH, CH₃)	NCP	89	4-chloro-3-methylphenol	98
			4,6-dichloro-3-methylphenol	2
1,3-dimethoxybenzene (OCH₃, OCH₃)	NCP	79	4-chloro-1,3-dimethoxybenzene	91
			4,6-dichloro-1,3-dimethoxy-benzene	9

TABLE I (cont'd.)

Structure	Reagent	Yield	Product	%
3-methoxytoluene (OCH₃, CH₃ meta)	NCP	85	4-chloro-3-methylmethoxybenzene	100
3,5-dimethylphenol (OH, CH₃, CH₃)	NCTA[e]	95	4-chloro-3,5-dimethylphenol	96
			4,6-dichloro-3,5-dimethylphenol	4

[a] Equimolar quanitities of substrate and N-chloro compound.
[b] Yield of products isolated from reaction, based on N-chloro compound.
[c] NCP = N-chloropiperidine.
[d] NCTA = N-chlorotriethylammonium chloride.[8]
[e] Two-fold excess of substrate, reaction temperature -17°C.

TABLE II

RELATIVE REACTIVITY AND SELECTIVITY OF N-CHLORINATED AMINES IN TRIFLUOROACETIC ACID

N-Chloro Compound	Reactivity Relative to NCP[a]	Ratio of 4- to 2-Chlorination[b]
CIN–NCl (piperazine)	160 000	6.0
O–NCl (morpholine)	17 000	20
Ph–CH$_2$N(CH$_3$)Cl	200	48
Cyclohexyl–N(CH$_3$)Cl	9	66
Piperidine–NCl	1	99
Quinuclidine $\overset{+}{N}$Cl Cl$^-$	0.2	500

[a]Determined from chlorination of 2-chloromethoxybenzene.
NCP = N-chloropiperidine.
[b]From the chlorination of methoxybenzene.

Unchecked Procedures

Accepted for checking during the period September 1, 1987 through August 1, 1988. An asterisk (*) indicates that the procedure has been subsequently checked.

In accordance with a policy adopted by the Board of Editors, beginning with Volume 50 and further modified subsequently, procedures received by the Secretary and subsequently accepted for checking will be made available upon request to the Secretary, if the request is accompanied by a stamped, self-addressed envelope. (Most manuscripts require 54¢ postage).

Address requests to:

> Professor Jeremiah P. Freeman
> Organic Syntheses, Inc.
> Department of Chemistry
> University of Notre Dame
> Notre Dame, Indiana 46556

It should be emphasized that the procedures which are being made available are unedited and have been reproduced just as they were first received from the submitters. There is no assurance that the procedures listed here will ultimately check in the form available, and some of them may be rejected for publication in *Organic Syntheses* during or after the checking process. For this reason, *Organic Syntheses* can provide no assurance whatsoever that the procedures will work as described and offers no comment as to what safety hazards may be involved. Consequently, more than usual caution should be employed in following the directions in the procedures.

Organic Syntheses welcomes, on a strictly voluntary basis, comments from persons who attempt to carry out the procedures. For this purpose, a Checker's Report form will be mailed out with each unchecked procedure ordered. Procedures which have been checked by or under the supervision of a member of the Board of Editors will continue to be published in the volumes of *Organic Syntheses*, as in the past. It is anticipated that many of the procedures in the list will be published (often in revised form) in *Organic Syntheses* in future volumes.

2423R*   Palladium Catalyzed Coupling of Vinyl Triflates with
         Organostannanes: 1-Vinyl-4-tert-Butylcyclohexene and 1-(4-tert-
         Butylcyclohex-1-enyl)propenone
         G. T. Crisp, W. J. Scott, and J. K. Stille, Department of Chemistry,
         Colorado State University, Ft. Collins, CO 80523

2443*   4,13-Diaza-18-crown-6
         V. J. Gatto, S. R. Miller, and G. W. Gokel, Department of Chemistry,
         University of Miami, Coral Gables, FL 33124

2451R   Ester Homologation via Ynolate Anions: Methyl 3-Cyclohexenylacetate
         C. J. Kowalski and S. Sakdarat, Synthetic Chemistry Department,
         Smith Kline and French Laboratories, P.O. Box 7929, Philadelphia, PA
         19101

2456R*   Intramolecular Cyclization of cis-cis-1,5-Cyclooctadiene Using
         Hypervalent Iodine: Synthesis of Bicyclo[3.3.0]octa-2,6-dione
         R. M. Moriarty, M. P. Ducan, O. Prakash, and R. K. Vaid, Department
         of Chemistry, University of Illinois at Chicago, Chicago, IL 60680

2463   Allyltributyltin
         N. G. Halligan and L. C. Blaszczak, Department of Chemical,
         Infectious Disease and Cancer Reserach, Lilly Corporate Center,
         Lilly Research Laboratories, Indianapolis, IN 46285

2465*   Cyclopentadiene Annulation via the Skattebøl Rearrangement: (1R)-
         9,9-Dimethyltricyclo[6.1.1.0^{2,6}]deca-2,5-diene
         L. A. Paquette and M. L. McLaughlin, Department of Chemistry, The
         Ohio State University, Columbus, OH 43210

2468R*   Synthesis of (Phenylthio)nitromethane
         A. G. M. Barrett, D. Dhanak, G. G. Graboski, and S. J. Taylor,
         Department of Chemistry, Northwestern University, Evanston, IL 60201

2469*   Yeast Reduction of 2,2-Dimethylcyclohexane-1,3-dione: (S)-(+)-3-
         Hydroxy-2,2-dimethylcyclohexanone
         K. Mori and H. Mori, Department of Agricultural Chemistry, The
         University of Tokyo, Bunkyo-Ku, Tokyo 113, Japan

2472   Propargylation of Alkyl Halides: Synthesis of (E)-6,10-Dimethyl-
         5,9-undecadien-1-yne and (E)-7,11-Dimethyl-6,10-dodecadien-2-yn-1-ol
         J. Hooz, J. Cabezas, S. Musmanni, and J. Calzada, Department of
         Chemistry, University of Alberta, Edmonton, Alberta, Canada T6G 2G2

2473*   Preparation of 4-(S)-Phenylmethyl-2-Oxazolidinone
         J. R. Gage and D. A. Evans, Department of Chemistry, Harvard
         University, Cambridge, MA 02138

2474*   A Highly Diastereoselective Aldol Reaction Using a Chiral
         Oxazolidinone Auxiliary
         J. R. Gage and D. A. Evans, Department of Chemistry, Harvard
         University, Cambridge, MA 02138

2476*   Immonium Ion Based Diels-Alder Reactions: N-Benzyl-2-azanorbornene
        P. A. Grieco and S. D. Larsen, Department of Chemistry, Indiana
        University, Bloomington, IN 47405

2477*   Transesterification of Methyl Esters of Aromatic and α,β-Unsaturated
        Acids with Bulky Alcohols: (-)-Menthyl Cinnamate and (-) Menthyl
        Nicotinate
        O. Meth-Cohn, Research and Development Director, Sterling Organics
        Ltd., Edgefield Avenue, Fawdon, Newcastle-on-tyne NE3 3TT, United
        Kingdom

2478*   p-tert-Butylcalix[4]arene
        C. D. Gutsche and M. Iqbal, Department of Chemistry, Washington
        University, St. Louis, MO 63130

2479   p-tert-Butylcalix[6]arene
        C. D. Gutsche, B. Dhawan, and M. Leonis, Department of Chemistry,
        Washington University, St. Louis, MO 63130

2480*   p-tert-Butylcalix[8]arene
        J. H. Munch and C. D. Gutsche, Department of Chemistry, Washington
        University, St. Louis, MO 63130

2481*   Synthesis of 2-(Phenylsulfonyl)-1,3-Cyclohexadiene
        J. E. Bäckvall, S. K. Juntunen, and O. S. Andell, Department of
        Organic Chemistry, University of Uppsala, Box 531, S-751 21 Uppsala,
        Sweden

2482*   Directed Homogeneous Hydrogenation: Anti-Methyl 3-Hydroxy-2-
        methylpentanoate
        J. M. Brown, P. L. Evans, and A. P. James, University of Oxford, The
        Dyson Perrins Laboratory, South Parks Road, Oxford QX1 3QY, England

2483   A General Synthesis of Cyclobutanones from Olefins and Tertiary
        Amides: 3-Hexylcyclobutanone
        C. Schmit, J. B. Falmagne, J. Escudero, H. Vanlierde and L. Ghosez,
        Laboratoire de Chimie Organique de Synthese, Universite Catholique
        de Louvain, Place L. Pasteur 1, B-1348, Louvain-La-Neuve, Belgium

2484   Synthesis of Racemic cis-4,5-Dihydro-4,5-dimethyl-3-methylene-2(3H)-
        furanone
        S. E. Drewes, R. F. A. Hoole, and S. D. Freese, Department of
        Chemistry, University of Natal, Pietermaritzburg 3200, South Africa

2486   6-Iodo-9-(β-D-ribofuranosyl)purine
        V. Nair, Department of Chemistry, The University of Iowa,
        Iowa City, IA 52242

2488*   Allylic Acetoxylation of Cycloalkenes
        A. Heumann, B. Akermark, S. Hanson, and T. Rein, Université d'Aix
        Marseille, Faculté de St.-Jérome, IPSOI F 13397 Marseille Cedex 13,
        France

2490   3-Butylcyclobutenone
        R. L. Danheiser, S. Savariar, and D. D. Cha, Department of
        Chemistry, Massachusetts Institute of Technology, Cambridge, MA
        02139

2491   L-Valinol, By LiAlH$_4$ Reduction
        G. A. Smith, G. Hart, S. Chemburkar, K. Rein, T. V. Anklekar, A. L.
        Smith, and R. E. Gawley, Department of Chemistry, P.O. Box 249118,
        University of Miami, Coral Gables, FL 33124

2494   Nucleophilic Hydroxymethylation of Carbonyl Compounds: 1-
        (Hydroxymethyl)cyclohexanol
        K. Tamao, N. Ishida, Y. Ito, and M. Kumada, Department of Synthetic
        Chemistry, Kyoto University, Sakyo-Ku, Kyoto 606, Japan

2495   Alkylidenation of Ester Carbonyl Groups: (2E,4Z)-4-Ethoxy-2,4-
        Decadiene
        K. Takai, Y. Kataoka, T. Okazoe, K. Oshima, and K. Utimoto,
        Department of Industrial Chemistry, Faculty of Engineering, Kyoto
        University, Yoshida, Kyoto 606, Japan

2497   Rearrangement of 4-Aryl-4-hydroxy-2,3-dialkoxycyclobutenediones
        to Annulated Hydroquinones and Quinones: 5,6-Diethoxybenzofuran-
        4,7-dione
        S. T. Perri, P. Rice, and H. W. Moore, Department of Chemistry,
        University of California, Irvine, CA 92717

2498   Synthesis of Alkyl Propanoates by Haloform Reaction of Trichloro
        Ketones: Preparation of Ethyl 3,3-Diethoxypropanoate
        L. F. Tietze, E. Voss, and U. Hartfiel, Institut für Organische
        Chemie, der Georg August Universität, Tammannstrasse 2, D-3400
        Göttingen, Federal Republic of Germany

CUMULATIVE AUTHOR INDEX
FOR VOLUMES 65, 66, AND 67

This index comprises the names of contributors to Volumes **65**, **66**, and **67** only. For authors to previous volumes, see Volume **64**, which covers Volumes **60** through **64**, and either indices in Collective Volumes I through VI or the single volume entitled *Organic Syntheses, Collective Volumes, I, II, III, IV, V, Cumulative Indices,* edited by R. L. Shriner and R. H. Shriner

Abrams, S. R., **66**, 127
Adger, B. M., **67**, 187
Akutagawa, S., **67**, 20, 33
Almond, M. R., **66**, 132
Anderson, F., **67**, 121
Aslamb, M., **65**, 90

Baba, S., **66**, 60
Bäckvall, J. E., **67**, 105
Bankston, **66**, 180
Bercaw, J. E., **65**, 42
Bergman, J., **65**, 146
Bergman, R. G., **65**, 42
Block, E., **65**, 90
Boger, D. L., **65**, 32, 98; **66**, 142
Boeckman, R. K., Jr., **66**, 194
Boes, M., **67**, 52, 60
Bou, A., **65**, 68
Braxmeier, H., **65**, 159
Brotherton, C. E., **65**, 32
Brown, C. A., **65**, 224
Büchi, G. H., **66**, 29
Buter, J., **65**, 140

Casy, G., **67**, 193
Celerier, J. P, **67**, 170
Chang, T. C. T., **66**, 95
Coghlan, M. J., **67**, 157
Corey, E. J., **65**, 166

Danheiser, R. L., **66**, 1, 8, 14
Davidsen, S. K., **65**, 119

Davis, F. A., **66**, 203
Deardorff, D. R., **67**, 114
Deloisy-Marchalant, E., **67**, 170
Dent, W., **67**, 133
Dickman, D. A., **67**, 52, 60
Dupuis, J., **65**, 236

Eliel, E. L., **65**, 215
Enders, D., **65**, 173, 183
Engler, A., **66**, 37

Fey, P., **65**, 173, 183
Fink, D. M., **66**, 1, 8, 14
Forth, M. A., **67**, 187
Fouquey, C., **67**, 1
Fowler, K. W., **65**, 108
Fox, C. M. J., **66**, 108
Frye, S. V., **65**, 215
Fujisawa, T., **65**, 116, 121
Fujita, T. **67**, 44
Furuta, K., **67**, 76

Gadwood, R. C., **67**, 210
Georg, G. I., **65**, 32
Giese, B., **65**, 236
Goel, O. P., **67**, 69
Goodwin, G. B. T., **65**, 1
Gross, A. W., **65**, 166

Haese, W., **65**, 26
Hanessian, S., **65**, 243
Hegedus, L. S., **65**, 140

Holmes, A. B., **65**, 52, 61
Hobbs, F. W., Jr., **66**, 52
Hormi, O., **66**, 173
Hudlicky, T., **67**, 121, 202
Hsiao, C.-N., **65**, 135
Huang, Z., **66**, 211

Imwinkelried, R., **65**, 230; **67**, 180
Iwanaga, K., **67**, 76

Jacques, J., **67**, 1
Jadhav, P. K., **65**, 224
Jones, G. E., **65**, 52

Katagiri, T., **67**, 44, 48
Kellogg, R. M., **65**, 150
Kendrick, D. A., **65**, 52
Kesten, S., **67**, 69
King, A. O., **66**, 67
Kipphardt, H., **65**, 173, 183
Koppenhoefer, B., **66**, 151, 160
Kraus, G. A., **67**, 202
Kresze, G., **65**, 159
Krolls, U., **67**, 69
Krolski, M. E., **67**, 202
Kuhlmann, H., **65**, 26
Kume, F., **65**, 215
Kumobayashi, H., **67**, 33
Kuwajima, I., **65**, 17; **66**, 43, 87

Labadie, J. W., **67**, 86
Larson, G. L., **67**, 125
Le Gal, J. Y., **65**, 47
Ley, S. V., **66**, 108
Lhommet, G., **67**, 170
Lin, H.-S., **67**, 157, 163
Lindsay Smith, J. R., **67**, 222
Lombardo, L., **65**, 81
Loudon, G. M., **66**, 132
Lynch, J. E., **65**, 215

Macdonald, J. .E., **66**, 194
Maitte, P., **67**, 170
Martin, S. F., **65**, 119
Maruoka, K., **66**, 185; **67**, 176
Masuda, T., **66**, 22
McGuire, M. A., **65**, 140
McKeer, L. C., **67**, 222
Meyers, A. I., **67**, 52, 60

Mickel, S. J., **65**, 135
Mieles, L. R., **67**, 125
Miller, M. J., **65**, 135
Miller, S. A., **67**, 210
Mimura, S., **66**, 22
Montes de Lopez-Cepero, I., **67**, 125
Mook, R., Jr., **66**, 75
Mukaiyama, T., **65**, 6
Mullican, M. D., **65**, 98
Munsterer, H., **65**, 159
Myles, D. C., **67**, 114

Nakai, S., **66**, 185
Nakamura, E., **65**, 17; **66**, 43
Nakamura, T., **67**, 98
Narasaka, K., **65**, 6, 12
Negishi, E.-i., **66**, 60, 67
Neveu, M., **67**, 205
Nicholas, K. M., **67**, 141
Nix, M., **65**, 236
Noyori, R., **67**, 20
Nugent, W. A., **66**, 52
Nyström, J. E., **67**, 105

Ochiai, H., **67**, 98
Okano, K., **66**, 14
Olomucki, M., **65**, 47
Ort, O., **65**, 203
Otsuka, S., **67**, 33
Oxley, P. W., **67**, 187

Padwa, A., **67**, 133
Panek, J. S., **66**, 142
Paquette, L. A., **67**, 149, 157, 163
Patterson, J. W., **67**, 193
Pauley, D., **67**, 121, 205
Pericas, M. A., **65**, 68
Perni, R. B., **66**, 194
Phillips, G. W., **65**, 119.
Poulter, C. D., **66**, 211

Rambaud, M., **66**, 220
Ranu, B. C., **67**, 205
Rein, T., **67**, 105
Renaldo, A. F., **67**, 86
Riera, A., **65**, 68
Rigby, H. L., **67**, 205
Rodgers, W. B., **65**, 108
Rosenblum, M., **66**, 95

Saha, M., **67**, 141
Sakane, S., **67**, 176
Sand, P., **65**, 146
Sasse, M. J., **67**, 187
Sato, T., **66**, 116, 121
Schiess, M., **65**, 230
Schultze, L. M., **65**, 140
Schurig, V., **66**, 151, 160
Seebach, D., **65**, 230; **67**, 180
Seidler, P. F., **65**, 42
Serratosa, F., **65**, 68
Shaw, A. C., **66**, 127
Sher, P. M., **66**, 75
Simms, N., **66**, 95
Smith, G. C., **65**, 1
Spitzner, D., **6 6**, 37
Sporikous, C. N., **65**, 61
Stetter, H., **65**, 26
Stier, M., **67**, 69
Stille, J. K., **67**, 86
Stimmel, J. B., **66**, 132
Stringer, O. D., **66**, 203
Stryker, J. M., **65**, 42
Suga, K., **67**, 44
Sy, J., **67**, 202
Szczepanski, S. W., **65**, 14

Takabe, K., **67**, 44, 48
Takahashi, T., **66**, 60, 67
Takaya, H., **67**, 20
Takeda, A., **66**, 22
Tamaru, Y., **67**, 98
Tanaka, J., **67**, 44, 48
Tani, K., **67**, 33
Taylor, J. M., **67**, 222
Taylor, R. J. K., **67**, 193
Thomas, A. J., **66**, 194
Thompson, E. A., **66**, 132
Threlkel, R. S., **65**, 42
Truesdale, L., **67**, 13
Tasi, Y.-M., **66**, 1, 8, 14
Tsuboi, S., **66**, 22

Urabe, H., **66**

Varghese, V., **67**, 141
Villieras, J., **66**, 220
Vishwakarma, L. C., **66**, 203
Vogel, D. E., **66**, 29

Walshe, N. D. A., **65**, 1
Watanabe, S., **67**, 44
Waykole, L., **67**, 149
Wester, R. T., **65**, 108
Woodward, F. E., **65**, 1
Woodside, A. B., **65**, 211

Yamada, T., **67**, 48
Yamagata, T., **67**, 33
Yamamoto, H., **66**, 185; **67**, 76, 176
Yasuda, M., **66**, 142
Yoshida, Z., **67**, 98

Ziegler, F. E., **65**, 108

CUMULATIVE SUBJECT INDEX FOR VOLUMES 65, 66 AND 67

This index comprises subject matter for Volumes **65**, **66**, and **67**. For subjects in previous volumes, see the cumulative indices in Volume **64**, which covers Volumes **60** through **64**, and either the indices in Collective Volumes I through VI or the single volume entitled *Organic Syntheses, Collective Volumes I, II, III, IV, V, Cumulative Indices*, edited by R. L. Shriner and R. H. Shriner.

The index lists the names of compounds in two forms. The first is the name used commonly in procedures. The second is the systematic name according to **Chemical Abstracts** nomenclature, accompanied by its registry number in parentheses. While the systematic name is indexed separately, it also accompanies the common name. Also included are general terms for classes of compounds, types of reactions, special apparatus, and unfamiliar methods.

Most chemicals used in the procedure will appear in the index as written in the text. There generally will be entries for all starting materials, reagents, intermediates, important by-products, and final products. Entries in capital letters indicate compounds, reactions, or methods appearing in the title of the preparation.

Acetamide, N,N-dimethyl-, **67**, 98

Acetamide, N-hydroxy-N-phenyl-, **67**, 187

Acetic acid, chloro-, 5-methyl-2-(1-methyl-1-phenylethyl)cyclohexyl ester,
[1R-(1α,2β,5α)]-, **65**, 203

Acetic acid, (diethoxyphosphinyl)-, ethyl ester; (867-13-0), **66**, 224

Acetic acid ethenyl ester, **65**, 135

Acetic acid, (hexahydro-2H-azepin-2-ylidene)-, ethyl ester, (Z)-, **67**, 170

Acetic acid, hydrazinoimino-, ethyl ester; (53085-26-0), **66**, 149

Acetic acid, lithium salt, dihydrate, **67**, 105

Acetic acid, palladium(2+) salt, **67**, 105

Acetic acid, trifluoro-, anhydride, **65**, 12

Acetic acid vinyl ester, **65**, 135

Acetohydroxamic acid, N-phenyl-, **67**, 187

ACETONE TRIMETHYLSILYL ENOL ETHER: SILANE, (ISOPROPENYLOXY)-
TRIMETHYL-; SILANE, TRIMETHYL[(1-METHYLETHENYL)OXY]-;
(1833-53-0), **65**, 1

Acetonitrile, purification, **66**, 101

Acetophenone; Ethanone, 1-phenyl-; (98-86-2), **65**, 6, 119

Acetophenone silyl enol ether: Silane, trimethyl[(1-phenylvinyl)oxy]-;
Silane, trimethyl[(1-phenylethenyl)oxy]-; (13735-81-4), **65**, 12

4-ACETOXYAZETIDIN-2-ONE: 2-AZETIDINONE, 4-HYDROXY-ACETATE (ESTER):
2-AZETIDINONE, 4-(ACETYLOXY)-; (28562-53-0), **65**, 135

1-ACETOXY-4-BENZYLAMINO-2-BUTENE: 2-BUTEN-1-OL, 4-CHLORO-,
ACETATE (E)-; (34414-28-3), **67**, 105

4-Acetoxy-3-chloro-1-butene: 3-Buten-1-ol, 2-chloro-, acetate; (96039-67-7), **67**, 105

1-ACETOXY-4-DIETHYLAMINO-2-BUTENE: 2-BUTEN-1-OL, 4-(DIETHYLAMINO)-,
ACETATE; (82736-47-8), **67**, 105

3-ACETOXY-5-HYDROXYCYCLOPENT-1-ENE, cis-: 4-CYCLOPENTEN-1,3-DIOL, MONOACETATE, cis-; (60410-18-6), **67**, 114

Acetylene; Ethyne; (74-86-2), **65**, 61

Acetylenes, arylsulfonyl-, **67** 149

α,β–Acetylenic esters from β–chloroalkylidene malonates, **66**, 178

3-ACETYL-4-HYDROXY-5,5-DIMETHYLFURAN-2(5H)-ONE, **66**, 108, 110

N-ACETYL-N-PHENYLHYDROXYLAMINE: ACETOHYDROXAMIC ACID, N-PHENYL; ACETAMIDE, N-HYDROXYL-N-PHENYL-; (1795-83-1), **67**, 187

Acrylonitrile; 2-Propenenitrile; (107-13-1), **65**, 236

Acyl chlorides, preparation, **66**, 116

Acylsilanes, preparation, **66**, 18

ADDITION OF TIN RADICALS TO TRIPLE BONDS, **66**, 75

Adipic acid monomethyl ester; (627-91-8), **66**, 116, 117, 120

Alane, (E)-1-decenyldiisobutyl-, **66**, 60, 61

(S)-Alanine (L-alanine); (56-41-7), **66**, 151-153, 159

Aldehydes, preparation from carboxylic acid, **66**, 124

Aldol reaction, **65**, 6, 12

Alkenylzinc reagents, **66**, 64

Alkylation, of ester enolate, **66**, 87

(R)-Alkyloxiranes, properties, **66**, 169

(R)-ALKYLOXIRANES OF HIGH ENANTIOMERIC PURITY, **66**, 160

ALKYNE ISOMERIZATION, **66**, 127

Allene, 1-methyl-1-(trimethylsilyl)-, **66**, 8, 9

Allenes, synthesis by Claisen rearrangement, **66**, 22

β–Allenic esters, conjugation by alumina, **66**, 26

ALLENYLSILANES, SYNTHESIS OF, **66**, 1

Allyl alcohol; (107-18-6), **66**, 14, 16, 21

ALLYLCARBAMATES, **65**, 159

(E)-3-Allyloxyacrylic acids, **66**, 33, 34

Allyl trimethylsilyl ether, metallation, **66**, 18

Aluminate(1-), dihydrobis-2-(methoxyethanalato)-, sodium, **67**, 13

Aluminate(1-), tetrahydro-, lithium, **67**, 69

Aluminum chloride, **66**, 90

Aluminum oxide, **66**, 23

Aluminum, hydrobis(2-methylpropyl)-; (1191-15-7), **66**, 193

Aluminum, triisobutyl, **67**, 176

Aluminum, tripropyl-; (102-67-0), **66**, 193

Amberlyst 15 ion exchange resin, **66**, 203, 206

Amino acids, determination of enantiomeric purity, **66**, 153

1-AMINO-2-METHOXYMETHYLPYRROLIDINE, (R)-(+)- (RAMP), **65**, 173

1-AMINO-2-METHOXYMETHYLPYRROLIDINE, (S)-(-)- (SAMP), **65**, 173, 183

3-Aminopropylamine: 1,3-Propanediamine; (109-76-2), **65**, 224

Ammonia, **66**, 133, 135

Anisole, 2,4-dichloro-, **67**, 222

[3+3] Annulation, **66**, 4, 8, 10, 11

Asymmetric alkylation, **67**, 60

Asymmetric synthesis, **65**, 183, 215

    of cycloalkane dicarboxylates, **67**, 76

AZA-ENE REACTION, **65**, 159

1H-Azepine, hexahydro-2-propyl-; (85028-29-1), **66**, 193

2H-Azepine, 3,4,5,6-tetrahydro-7-methoxy-, **67**, 170

2-AZETIDINONE, 4-(ACETYLOXY)-, **65**, 135

2-AZETIDINONE, 4-HYDROXY-ACETATE (ESTER), **65**, 135

2-AZETIDINONE, 3-METHOXY-1,3-DIMETHYL-4-PHENYL-, **65**, 140

Azobisisobutyronitrile (AIBN); (78-67-1), **66**, 77, 79, 86; **67**, 86

AZOMETHINE YLIDE EQUIVALENT, **67**, 133

Beckmann rearrangement, **66**, 185, 189, 190

Benzaldehyde; (100-52-7), **65**, 119; **66**, 203

Benzaldehyde, 2-bromo-4,5-dimethoxy-, **65**, 108

Benzaldehyde, 3-nitro-, **67**, 180

Benzeneamine, N-hydroxy-, **67**, 187

Benzeneamine, 2-methyl-3-nitro-, **65**, 146

Benzene, (1-cyclopenten-1-ylsulfonyl)-, **67**, 157, 163

Benzene, 2,4-dichloro-1-methoxy-, **67**, 222

1,2-Benzenediol; (120-80-9), **66**, 184; **67**, 98

Benzene, 1-(ethynylsulfonyl)-4-methyl-, **67**, 149

Benzenemethanamine, N-(methoxymethyl)-N-[(trimethylsilyl)methyl]-, **67**, 133

Benzenemethanamine, N-[(trimethylsilyl)methyl]-, **67**, 133

Benzenemethanol, α-methyl-3-nitro-, **67**, 180

Benzenemethanethiol, **65**, 215

Benzeneseleninic acid, **67**, 157

Benzenesulfonamide, **66**, 203

Benzenesulfonic acid, hydrazide, **67**, 157

Benzenesulfonoselenoic acid, Se-phenyl ester, **67**, 157

1-(BENZENESULFONYL)CYCLOPENTENE: BENZENE, (1-CYCLOPENTEN-

1-YLSULFONYL)-; (64740-90-5), **67**, 157, 163

Benzenesulfonyl hydrazide: Benzenesulfonic acid, hydrazide; (80-17-1), **67**, 157

2H-1-BENZOPYRAN-3-CARBOXYLIC ACID, 5,6,7,8-TETRAHYDRO-2-OXO-,

METHYL ESTER, **65**, 98

p-Benzoquinone; (106-51-4), **67**, 105

4H-1,3-BENZOXATHIIN, HEXAHYDRO-4,4,7-TRIMETHYL-, **65**, 215

Benzoyl chloride, 4-nitro-, **67**, 86

Benzyl alcohol, α-methyl-m-nitro-, **67**, 180

Benzylamine; (100-46-9), **67**, 105, 133

Benzylchlorobis(triphenylphosphine)-, trans-: Palladium, benzylchlorobis-
(triphenylphosphine)-, trans-; Palladium, chloro(phenylmethyl)bis(triphenylphosphine)-, (SP-4-3)-; (22784-59-4), **67**, 86

3-Benzyl-5-(2-hydroxyethyl)-4-methyl-1,3-thiazolium chloride; (4568-71-2), **65**, 26

N-Benzylidenebenzenesulfonamide; (13909-34-7), **66**, 203, 204, 206, 210

Benzylidenemalononitrile: Malononitrile, benzylidene-; Propanedinitrile,
(phenylmethylene)-; (2700-22-3), **65**, 32

Benzyl mercaptan: α–Toluenethiol; Benzenemethanethiol; (100-53-8), **65**, 215

N-BENZYL-N-METHOXYMETHYL-N-(TRIMETHYLSILYL)METHYLAMINE:
BENZENEMETHANAMINE, N-(METHOXYMETHYL)-N-[(TRIMETHYL-
SILYL)METHYL]-; (93102-05-7), **67**, 133

Benzyltriethylammonium chloride, **66**, 204

N-Benzyl-N-(trimethylsilyl)methylamine: Benzenemethanamine, N-[(trimethylsilyl)methyl]-; (53215-95-5), **67**, 133

BETA-LACTAM, **65**, 135, 140

BIARYL SYNTHESIS, **66**, 67

Biaryls, preparation by Pd-catalyzed aryl-aryl coupling, **66**, 73

BICYCLO[3.1.1]HEPTANE, 6,6-DIMETHYL-2-METHYLENE-, (1S)-, **65**, 224

BICYCLO[3.1.1]HEPT-2-ENE, 2,6,6-TRIMETHYL-, (1S)-, **65**, 224

BICYCLO[4.3.0]NON-1-EN-4-ONE: 5H-INDEN-5-ONE, 1,2,3,3a,4,6-
HEXAHYDRO-, **67**, 163

Bicyclo[2.2.2]octan-2-ones, preparation, **66**, 37, 40

BINAP, (R)-(+)- and (S)-(-)-, **67**, 20, 33

1,1'-Binaphthalene, 2,2'-dibromo-, **67**, 20

1,1'-BINAPHTHALENE-2,2'-DIOL, (R)-(+)-: [1,1'-BINAPHTHALENE]-2,2'-DIOL, (R)-; (18531-94-7), **67**, 13

Binaphthol, (±)-; (41024-90-2), (602-09-5), **67**, 1, 13, 20

Binaphthol, R-(+)-; (18531-94-7); S-(-)-; (1853-99-2), **67**, 13

1,1'-BINAPHTHYL-2,2'-DIYL HYDROGEN PHOSPHATE, (R)-(-)-: R(-) DI-NAPHTHO[2,1-d:1'2'-f][1,3,2]DIOXAPHOSPHEPIN, 4-HYDROXY-4-OXIDE; (39648-67-4), **67**, 1, 13

1,1'-BINAPHTHYL-2,2'-DIYL HYDROGEN PHOSPHATE, (S)-(+)-: S(+) DI-NAPHTHO[2,1-d:1'2'-f][1,3,2]DIOXAPHOSPHEPIN, 4-HYDROXY-4-OXIDE; (35193-64-7), **67**, 1

BINAPO, (R)-(+)-, and (S)-(-)-, **67**, 20

[1,1'-BIPHENYL]-2,2'-DICARBOXALDEHYDE, 4,4',5,5'-TETRAMETHOXY-, **65**, 108

Biphenyl, 2-methyl-4'-nitro-; (33350-73-1), **66**, 74

[2,2'-BIS(DIPHENYLPHOSPHINO)-1,1'-BINAPHTHYL (BINAP), (R)-(+)- and (S)-(-)-: PHOSPHINE, [1,1'-BINAPHTHALENE]-2,2'-DIYLBIS-[DIPHENYL-, (R)- and (S)-]; (R)-: (76189-55-4), (S)-; (76189-56-5), **67**, 20, 33

2,2'-Bis(diphenylphosphino)-1,1'-binaphthyl-(η^4-1,5-cyclooctadiene)-rhodium(I) perchlorate, (+)- and (-): Rhodium(1+), [[1,1'-binaph-thalene]-2,2'diylbis[diphenylphosphine]-P,P'][(1,2,5,6-η)–1,5-cyclo-octadiene]-, stereoisomer perchlorate; (+)-; (82822-45-5); (-)-; (82889-98-3), **67**, 33

2,2'-Bis(diphenylphosphinyl)-1,1'-binaphthyl, (±)-, (S)-(-)-, and (R)-(+)- [(±)-, (S)-(-)-, and (R)-(+)-BINAPO]: Phosphine oxide, [1,1'-binaphthalene]-2,2'-diylbis[diphenyl-, (±)-, (S)-, and (R)-]; (±)-; (866632-33-9); (S)-; (94041-18-6); (R)-; (94041-16-4), **67**, 20

1,2-Bis(tributylstannyl)ethylene, (E)-: Stannane, vinylenebis[tributyl-, (E)-;
    Stannane, 1,2-ethenediylbis[dibutyl-, (E)-; (14275-61-7), **67**, 86
1,2-Bis(trimethylsiloxy)cyclobut-1-ene: Silane, (1-cyclobuten-1,2-
    ylenedioxy)bis[trimethyl-; Silane, [1-cyclobutene-1,2-
    diylbis(oxy)]bis[trimethyl-; (17082-61-0), **65**, 17
1,4-BIS(TRIMETHYLSILYL)BUTA-1,3-DIYNE: 2,7-DISILAOCTA-3,5-DIYNE,
    2,2,7,7-TETRAMETHYL-; SILANE, 1,3-BUTADIYNE-1,4-DIYLBIS-
    [TRIMETHYL-; (4526-07-2), **65**, 52
BNP acid, R(-) : R(-) Dinaphtho[2,1-d:1'2'-f][1,3,2]dioxaphosphepin,
    4-hydroxy-4-oxide, **67**, 1
    Compound with 8α, 9–cinchonan-9-ol; (40481-36-5), **67**, 1
BNP acid, S(+) : S(+) Dinaphtho[2,1-d:1'2'-f][1,3,2]dioxaphosphepin,
    4-hydroxy-4-oxide, **67**, 1
    Compound with 9S-cinchonan-9-ol; (3974950-3), **67**, 1
Borate(1-), tetrafluoro-, hydrogen, compound with oxybis[methane], **67**, 141
Boron trifluoride etherate: Ethyl ether, compd. with boron fluoride (BF_3) (1:1);
    Ethane, 1,1'-oxybis-, compd. with trifluoroborane (1:1); (109-63-7), **65**, 17l;
    67, 52, 205
4-Bromobutanenitrile: Butyronitrile, 4-bromo-; Butanenitrile, 4-bromo-;
    (5332-06-9), **67**, 193
2-Bromo-2-butene (cis and trans mixture): 2-Butene, 2-bromo-; (13294-71-8), **65**, 42
1-Bromo-3-chloro-2,2-dimethoxypropane: 2-Propanone, 1-bromo-3-chloro-,
    dimethyl acetal; Propane, 1-bromo-3-chloro-2,2-dimethoxy-; (22089-54-9),
    65, 2
1-Bromo-1-chloromethane, **67**, 76
6-Bromo-3,4-dimethoxybenzaldehyde: Benzaldehyde, 2-bromo-4,5-dimethoxy-;
    (5392-10-9), **65**, 108

6-Bromo-3,4-dimethoxybenzaldehyde cyclohexylimine: Cyclohexanamine
N-[(2-bromo-4,5-dimethoxyphenyl)methylene]-; (73252-55-8), **65**, 108

1-BROMO-1-ETHOXYCYCLOPROPANE: CYCLOPROPANE, 1-BROMO-1-
ETHOXY-; (95631-62-2), **67**, 210

BROMOMETHANESULFONYL BROMIDE; METHANESULFONYL BROMIDE,
BROMO-; (54730-18-6), **65**, 90

2-(Bromomethyl)-2-(chloromethyl)-1,3-dioxane: 1,3-Dioxane, 2-(bromomethyl)-
2-(chloromethyl)-; (60935-30-0), **65**, 32

N-Bromomethylphthalimide: Phthalimide, N-(bromomethyl)-; 1H-Isoindole-
1,3-(2H)-dione, 2-(bromomethyl)-; (5332-26-3), **65**, 119

1-Bromo-4-nitrobenzene, **66**, 68

N-BROMOSUCCINIMIDE: SUCCINIMIDE, N-BROMO-; 2,5-PYRROLIDINEDIONE,
1-BROMO-; (128-08-5), **65**, 243

o-Bromotoluene, **66**, 69

Brook rearrangement, **66**, 19

Butadiyne, **65**, 52

Butane, 4-bromo-, 1,1,1-trimethoxy-, **67**, 193

Butane, 1-chloro-, **65**, 61

Butanedioic acid, 2,3-bis(benzoyloxy)-, [R-(R*,R*)]- and [S-(R*,R*)]-, **67**, 20

Butanedioic acid, bis[5-methyl-2-(1-methylethyl)-cyclohexyl]ester,
[1R-[1α(1R*,2S*,5R*), 2β,5α]-, **67**, 76

Butanedioic acid, dioxo-, diethyl ester; (59743-08-7), **66**, 149

Butanedioic acid, tetrahydroxy-, disodium salt; (866-17-1), **66**, 149

Butanenitrile, 4-bromo-, **67**, 193

Butanoic acid, 4-bromo-, ethyl ester, **67**, 98

Butanoic acid, 4-iodo-, ethyl ester, **67**, 98

1-Butanol, 2-amino-3-methyl-, (S)-, **67**, 52

1-Butanone, 3-hydroxy-3-methyl-1-phenyl-, **65**, 6, 12

2-Butene, 2-bromo-, **65**, 42

2-Butene, 1-bromo-3-methyl-; (870-63-3), **66**, 85

2-Butene, 2-methyl-, **65**, 159

2-Butenoic acid, 2-methyl-, methyl ester, (E); (6622-76-0), **66**, 94

2-Butenoic acid, 4-(4-nitrophenyl)-4-oxo-, ethyl ester, **67**, 86

2-Buten-1-ol, 4-chloro-, acetate (E)-, **67**, 105

3-Buten-1-ol, 2-chloro-, acetate, **67**, 105

2-Butenol, 4-(diethylamino)-, acetate, **67**, 105

2-Buten-1-one, 3-methyl-1-phenyl-, **65**, 12

3-Buten-2-one; (78-94-4), **65**, 26

N-tert-BUTOXYCARBONYL-L-LEUCINAL: CARBAMIC ACID, (1-FORMYL-3-
METHYLBUTYL)-, 1,1-DIMETHYLETHYL ESTER, (S)-; (5821-45-2), **67**, 69

N-tert-Butoxycarbonyl-L-leucine N-methyl-O-methylcarboxamide: Carbamic acid,
[1-[(methoxymethylamino)carbonyl]-3-methylbutyl]-, 1,1-dimethylethyl ester,
(S)-; (87694-50-6), **67**, 69

N-tert-Butoxy-L-leucine hydrate: Leucine, N-carboxy-, N-tert-butyl ester, L-;
L-Leucine, N-[(1,1-dimethylethoxy)carbonyl]-; (13139-15-6), **67**, 69

tert-Butyl alcohol, potassium salt, **67**, 125

tert-BUTYL ACETOTHIOACETATE; (15925-47-0), **66**, 108, 112, 114

Butyl bromide, **66**, 118

tert-Butylhydrazine hydrochloride: Hydrazine, tert-butyl, monohydrochloride;
Hydrazine, (1,1-dimethylethyl)-, monohydrochloride; (7400-27-3), **65**, 166

3-tert-Butyl-4-hydroxy-5-methylphenyl sulfide, **66**, 16

tert-Butyllithium, **66**, 15, 16, 18, 67, 69; **67**, 60, 210

Butyllithium: Lithium, butyl-; (109-72-8), **65**, 98, 108, 119;
66, 14, 16, 37, 39, 88, 194, 196; **67**, 44, 48

Butylmagnesium bromide, **66**, 116, 118

N-tert-Butyl-N-tert-octyl-O-tert-butylhydroxylamine: 2-Pentanamine, N-(1,1-dimethylethoxy)-N-(1,1-dimethylethyl)-2,4,4-trimethyl-; (90545-93-0), **65**, 166

TERT-BUTYL-TERT-OCTYLAMINE; 2-PENTANAMINE, N-(1,1-DIMETHYLETHYL)-2,4,4-TRIMETHYL-; (90545-94-1), **65**, 166

S-tert-Butyl 3-oxobutanethioate, **66**, 108, 109

2-BUTYNOIC ACID, 4-CHLORO-, METHYL ESTER, **65**, 47

3-Butyn-2-ol; (65337-13-5), **67**, 141

Carbamic acid, dichloro-, methyl ester, **65**, 159

Carbamic acid, (1-formyl-3-methylbutyl)-, 1,1-dimethylethyl ester, (S)-, **67**, 69

Carbamic acid, [1-[(methoxymethylamino)carbonyl]-3-methylbutyl]-, 1,1-dimethylethyl ester, (S)-, **67**, 69

CARBAMIC ACID, (2-METHYL-2-BUTENYL)-, METHYL ESTER, **65**, 159

Carbamic acid, methyl ester, **65**, 159

Carboalumination of alkynes, **66**, 64

Carbocupration of alkynes, **66**, 65

1-Carbomethoxy-1-methylethyl-3-oxobutanoate, **66**, 110

2-CARBOMETHOXY-3-VINYLCYCLOPENTANONE; (75351-19-8), **66**, 52, 53, 56, 59

Carbonic acid, dicesium salt, **65**, 150

Carbonochloridic acid, methyl ester, **65**, 47

Carboxylic acids, reduction, **66**, 160

(E)-(Carboxyvinyl)trimethylammonium betaine; (54299-83-1), **66**, 29, 30, 33, 36

(R)-(-)-Carvone; (6485-40-1), **66**, 8, 9, 13

CATALYTIC TRANSFER HYDROGENATION, **67**, 187

Catechol, **66**, 180, 182

Cellulose chromatography, **66**, 213, 216

Cerate(2-), hexanitrato-, diammonium, **67**, 141

Ceric ammonium nitrate: Cerate(2-), hexanitrato-, diammonium; Cerate(2-), hexakis(nitrato-O)-, diammonium, (OC-6-11); (16774-21-3), **67**, 141

Cesium carbonate: Carbonic acid, dicesium salt; (534-17-8), **65**, 150

CESIUM THIOLATES, **65**, 150

Chiral auxiliary, **65**, 173, 183, 203, 215

Chirasil-Val, **66**, 153, 154

Chloroacetaldehyde diethyl acetal; (621-62-5), **66**, 96, 100, 107

CHLORINATION OF ELECTRON-RICH BENZENOI COMPOUNDS, **67**, 222

(S)-2-Chloroalkanoic acids

    determination of enantiomeric purity, **66**, 154

    properties, **66**, 155

(S)-2-CHLOROALKANOIC ACIDS OF HIGH ENANTIOMERIC PURITY, **66**, 151

(S)-2-CHLOROALKANOIC ACIDS VIA (S)-2-CHLORO-1-ALKANOLS, **66**, 160

(S)-2-Chloroalkan-1-ols, properties, **66**, 168

N-Chloroamines, **67**, 222

1-Chlorobutane: Butane, 1-chloro-; (109-69-3), **65**, 61

α–Chloroenamines, **66**, 119

2-Chloroethanol; Ethanol, 2-chloro-; (107-07-3), **65**, 150

2-Chloroethyl dichlorophosphate: Phosphorodichloridic acid, 2-chloroethyl ester; (1455-05-6), **65**, 68

Chlorohydrins, preparation from α–chloro acids, **66**, 161

2-Chloromethoxybenzene: Anisole, o-chloro-; Benzene, 1-chloro-2-methoxy-; (766-51-8), **67**, 222

(S)-2-Chloro-3-methylbutan-1-ol, **66**, 164

6-Chloromethyl-2,2-dimethyl-1,3-dioxen-4-one; (81956-31-2), **66**, 194, 195, 202

(2S,3S)-2-Chloro-3-methylpentanoic acid, **66**, 154

(2S,3S)-2-Chloro-3-methylpentan-1-ol, **66**, 164

(S)-2-Chloro-4-methylpentan-1-ol, **66**, 164

Chloromethyltrimethylsilane **67,** 133

N-Chloromorpholine: Morpholine, 4-chloro-; (23328-69-0), **67**, 222

m-Chloroperoxybenzoic acid, **66**, 204

(S)-2-CHLOROPROPANOIC ACID; (29617-66-1), **66**, 151, 154, 159, 160, 172

(S)-2-Chloropropan-1-ol; (19210-21-0), **66**, 160, 164, 172

N-Chlorosuccinimide; (128-09-6), **66**, 211, 214, 219

Chlorosulfonyl isocyanate: Sulfuryl chloride isocyanate; (1189-71-5), **65**, 135

Chlorotetrolic ester, **65**, 47

1-Chloro-N,N,2-trimethylpropenylamine; (26189-59-3), **66**, 116, 117, 119, 120

Chlorotrimethylsilane: Silane, chlorotrimethyl-; (75-77-4), **65**, 1, 6, 61;

    66, 14, 16, 21, 44

    as catalyst for conjugate addition, **66**, 47

1-Chloro-3-trimethylsilylpropane; (2344-83-4), **66**, 88, 90, 94

Chromium carbonyl, **65**, 140

Chromium hexacarbonyl: Chromium carbonyl (OC-6-11); (13007-92-6), **65**, 140

Chromium pentacarbonyl (1-methoxyethylidene)-, **65**, 140

Cinchonidine: 8α, 9R-Cinchonan-9-ol; (485-71-2), **67**, 1

Cinchonine: 9S-Cinchonan-9-ol; (119-10-5), **67**, 1

Cinnamyl alcohol; (104-54-1), **66**, 30, 32, 36

CITRONELLAL, (R)-(+)-: 6-OCTENAL, 3,7-DIMETHYL, (R)-(+)-; (2385-77-5),

    67, 33

Claisen rearrangement, of propargyl alcohols, **66**, 22

Cobalt(1+), hexacarbonyl[μ[(2,3-η:2,3-η)-1-methyl-2-propynylium]]di-, tetra-

    fluoroborate(1-), **67**, 141

Cobalt, octacarbonyldi-, **67**, 141

CONJUGATE ADDITION/CYCLIZATION OF A CYANOCUPRATE, **66**, 52

CONJUGATE ADDITION, ZINC HOMOENOLATES TO ENONES, **66**, 43, 50

CONJUGATED DIENES, SYNTHESIS OF, **66**, 60, 64

Copper(I) bromide (7787-70-4), **65**, 203; **66**, 2, 3, 44, 45

Copper, bromo[thiobis[methane]]-; (54678-23-8), **66**, 51

Copper chloride (CuCl); (7758-89-6), **66**, 180, 182, 184; **67**, 121

Copper(I) chloride - tetramethylethylenediamine complex, **65**, 52

Copper, compound with zinc (1:1), **67**, 98

Copper cyanide (CuCN), **66**, 53, 55

Copper iodide (CuI), **66**, 97, 100, 116, 118

Crotonaldehyde; (123-73-9), **67**, 210

Cuprate addition, **66**, 52, 95

Cuprate, dibutyl-, lithium, reaction with acetylene, **66**, 62

Cuprates, higher order, **66**, 57

Cuprous bromide/dimethyl sulfide; (54678-23-8), **66**, 51

Cuprous chloride-pyridine complex, **66**, 182

[2 + 2] CYCLOADDITION, **65**, 135

Cyclobutanecarboxamide; (1503-98-6), **66**, 132, 134, 136, 141

Cyclobutanecarboxylic acid; (3721-95-7), **66**, 133, 135, 141

Cyclobutanone, 2-(1-propenyl)-, (E)-, **67**, 210

CYCLOBUTANONES, **67**, 210

CYCLOBUTYLAMINE, **66**, 132

Cyclobutylamine hydrochloride; (6291-01-6), **66**, 134, 135, 141

Cyclohexanamine, **65**, 108

Cyclohexanamine, N-[(2-bromo-4,5-dimethoxyphenyl)methylene]-, **65**, 108

Cyclohexanamine, N-[(2-iodo-4,5-dimethoxyphenyl)methylene]-, **65**, 108

Cyclohexane, 1,1-diethoxy-, **65**, 17

CYCLOHEXANE, 1,2-BIS(METHYLENE)-, **65**, 90

CYCLOHEXANE, 1,2-DIMETHYLENE, **65**, 90

CYCLOHEXANE, 5-METHYL-1-METHYLENE-2-(1'-METHYLETHYL)-, R,R-, **65**, 81

CYCLOHEXANEBUTANOIC ACID, γ-OXO-, ETHYL ESTER, **65**, 17

Cyclohexanol, 2-(1-mercapto-1-methylethyl)-5-methyl-, [(1R-(1α,2α,5α)]-, **65**, 215

Cyclohexanol, 5-methyl-2-(1-methylethyl)-, [1R-(1α,2β,5α)]-, **67**, 76

Cyclohexanol, 5-methyl-2-(1-methylethyl)-, [1S-(1α,2β,5β)], **65**, 81

CYCLOHEXANOL, 5-METHYL-2-(1-METHYL-1-PHENYLETHYL)-, [1R-(1α,2β,5α)]-, **65**, 203

Cyclohexanone; (108-94-1), **65**, 98; **67**, 141

Cyclohexanone diethyl ketal; Cyclohexane, 1,1-diethoxy-; (1670-47-9), **65**, 17

Cyclohexanone, 5-methyl-2-(1-methylethyl)-, (2R-cis)-, **65**, 81

CYCLOHEXANONE, 5-METHYL-2-(1-METHYLETHYLIDENE)-, (R)-, **65**, 203, 215

Cyclohexanone, 5-methyl-2-(1-methyl-1-phenylethyl)-, (2R-trans)-, **65**, 203

Cyclohexanone, 5-methyl-2-(1-methyl-1-phenylethyl)-, (2S-cis)-, **65**, 203

Cyclohexanone, 5-methyl-2-[1-methyl-1-(phenylmethylthio)ethyl]-, (2R-trans)-; (79563-58-9); (2S-cis)-; (79618-04-5), **65**, 215

Cyclohexanone oxime, (100-64-1), **65**, 185, 187, 193

Cyclohexanone oxime methanesulfonate; (80053-69-6), **66**, 185, 186, 193

1-Cyclohexene-1-carboxaldehyde, 3-hydroxy-, **67**, 205

1-Cyclohexene-1-methanol, 4-(1-methyethenyl)-, **67**, 176

1-Cyclohexene-1-methanol, 4-(1-methylcyclopropyl)-, **67**, 176

Cyclohexene, 1-methyl-, **65**, 90

2-Cyclohexen-1-ol, 1-(1,3-dithian-2-yl)-, **67**, 205

2-Cyclohexen-1-one; (930-68-7), **66**, 44, 45, 51, 95, 97, 101, 107; **67**, 205

2-Cyclohexene-1-propanoic acid, 3-[(trimethylsilyl)oxy]-,
 ethyl ester; (90147-64-1), **66**, 51

Cyclohexylamine; Cyclohexanamine; (108-91-8), **65**, 108

1,3-CYCLOPENTADIENE, 1,2,3,4,5-PENTAMETHYL-, **65**, 42

Cyclopentadienes, synthesis, **66**, 11

Cyclopentadiene monoepoxide: 6-Oxabicyclo[3.1.0]hex-2-ene; (7129-41-1),
 67, 114

Cyclopentanes, synthesis, **66**, 10

CYCLOPENTANONE SYNTHESIS, **66**, 87, 92, 93

Cyclopentanone, 2-carbomethoxy-3-vinyl, **66**, 56

Cyclopentanone, 2-ethenyl-2-methyl; (88729-76-4), **66**, 94

Cyclopentene; (142-29-0), **67**, 157

4-Cyclopentene-1,3-diol, monoacetate, cis-, **67**, 114

2-Cyclopenten-1-one, 4,4-dimethyl-, **67**, 121, 205

2-CYCLOPENTEN-1-ONE, 3-METHYL-2-PENTYL-, **65**, 26

CYCLOPROPANATION, **67**, 176

Cyclopropane, 1-bromo-1-ethoxy-, **67**, 210

CYCLOPROPANE -1,2-DICARBOXYLIC ACID, (+)-(1S,2S)-: 1,2-CYCLOPRO-
 PANEDICARBOXYLIC ACID, (S,S)-(+)-; 1,2-CYCLOPROPANEDI-
 CARBOXYLIC ACID, (1S-trans)-; (14590-54-6) **67**, 76

1,2-Cyclopropanedicarboxylic acid, (S,S)-(+)- **67**, 76

1,2-Cyclopropanedicarboxylic acid, bis[5-methyl-2-(1-methylethyl)cyclohexyl]
 ester, **67**, 76

Cyclopropane, 1-trimethylsiloxy-1-ethoxy-, **66**, 44

Cyclopropanecarboxylic acid chloride, **66**, 176

CYCLOPROPENONE 1,3-PROPANEDIOL KETAL: 4,8-DIOXASPIRO[2.5]OCT-1-ENE;
 (60935-21-9), **65**, 32

Cyclopropylpropiolic acid ethyl ester, **66**, 177

Davis reagent, **66**, 203

Decanoic acid, ethyl ester, **67**, 125

Decanoic acid, 2-(methyldiphenylsilyl)-, ethyl ester, **67**, 125

Decanoic acid, 6-oxo-; (4144-60-9), **66**, 126

Decanoic acid, 6-oxo-, methyl ester; (61820-00-6), **66**, 120

(E)-1-Decenyldiisobutylalane, **66**, 60, 61

1-Decyne; (764-93-2), **66**, 60, 61, 66

2-DECYN-1-OL; (4117-14-0), **66**, 127, 128, 131

9-DECYN-1-OL; (17643-36-6), **66**, 127, 128, 131

1-DEOXY-2,3,4,6-TETRA-O-ACETYL-1-(2-CYANOETHYL)-α-D-GLUCOPYRANOSE:
D-GLYCERO-D-IDO-NONONONITRILE, 4,8-ANHYDRO-2,3-DIDEOXY-,
5,6,7,9-TETRAACETATE; (86563-27-1), **65**, 236

(I,I-Diacetoxyiodo)benzene, **66**, 136

DIALKOXYACETYLENES, **65**, 68

1,3-Diaminopropane; (109-76-2), **66**, 127, 128, 131

Diazotization of amino acids, **66**, 151, 156

2,3-O-Dibenzoyl-L-tartaric acid monohydrate, (-)- and (+)- [(-)-DBT and (+)-DBT monohydrate]: Butanedioic acid, 2,3-bis(benzoyloxy)-, [R-(R*,R*)]- and [S-(R*,R*)]-: [R-(R*,R*)]-; (2743-38-6); [S-(R*,R*)]-; (17026-42-5) **67**, 20

2,2'-Dibromo-1,1'-binaphthyl: 1,1'-Binaphthalene, 2,2'-dibromo-; (74866-28-7), **67**, 20

Dibromomethane: Methane, dibromo-; (74-95-3), **65**, 81

1,2-Di-tert-butoxy-1-chloroethene, (E)-: Propane, 2,2'-[(1-chloro-1,2-ethenediyl)bis(oxy)]bis[2-methyl-, (E); (70525-93-8), **65**, 58

1,2-Di-tert-butoxy-1,2-dichloroethane, dl-: Propane, 2,2'-[(1,2-dichloro-1,2-
ethanediyl)]bis(oxy)bis[2-methyl-, (R*,R*)-(±)-; (68470-80-4), **65**, 68

1,2-Di-tert-butoxy-1,2-dichloroethane, meso-: Propane, 2,2'-[(1,2-dichloro-
1,2-ethanediyl)bis(oxy)]bis[2-methyl-, (R*,S*)-; (68470-81-5), **65**, 68

2,3-Di-tert-butoxy-1,4-dioxane, cis-: 1,4-Dioxane, 2,3-bis(1,1-
dimethylethoxy)-, cis-; (68470-78-0), **65**, 68

2,3-Di-tert-butoxy-1,4-dioxane, trans-: 1,4-Dioxane,
2,3-bis(1,1-dimethylethoxy)-, trans-; (68470-79-1), **65**, 68

DI-TERT-BUTOXYETHYNE; PROPANE, 2,2'-[1,2-ETHYNEDIYLBIS(OXY)]BIS[2-
METHYL-; (66478-63-5), **65**, 68

Dicarbonyl(cyclopentadienyl)diiron, [(CO)$_2$CpFe]$_2$; (12154-95-9),
66, 96, 99, 106

Dicarbonyl(cyclopentadienyl)(ethyl vinyl ether)iron tetrafluoroborate,
66, 96, 106

Dicarbonyl(cyclopentadienyl)(trans-3-methyl-2-vinylcyclohexanone)iron
tetrafluoroborate, **66**, 97

Di-μ-chlorobis(η^4-1,5-cyclooctadiene)dirhodium(I): Rhodium, di-μ-chlorobis-
[(1,2,5,6-η)-1,5-cyclooctadiene]di-; (12092-47-6), **67**, 33

2,3-Dichloro-1,4-dioxane, trans-: 1,4-Dioxane, 2,3-dichloro-, trans-;
(3883-43-0), **65**, 68

(l,l-Dichloroiodo)benzene, **66**, 137

2,4-DICHLOROMETHOXYBENZENE: ANISOLE, 2,4-DICHLORO-; BENZENE,
2,4-DICHLORO-1-METHOXY-; (553-82-2), **67** 222

Dicobalt octacarbonyl: Cobalt, octacarbonyldi-, (Co-Co); (15226-74-1), **67**,
141

5,5-Dicyano-4-phenylcyclopent-2-enone 1,3-propanediol ketal:

    6,10-Dioxaspiro[4,5]dec-3-ene-1,1-dicarbonitrile, 2-phenyl-;

    (88442-12-0), **65**, 32

Dicyclopentadiene, **66**, 99

Dieckmann cyclization, **66**, 52

DIELS-ALDER REACTION, INVERSE ELECTRON DEMAND, **66**, 142, 147, 148

Diels-Alder reaction of triethyl 1,2,4-triazine-3,5,6-tricarboxylate, **66**, 150

Diels-Alder reactions, **66**, 40

Diethylamine, **66**, 145; **67**, 44, 48, 105

Diethyl aminomethylphosphonate: Phosphonic acid, (aminomethyl)-,

    diethyl ester; (50917-72-1), **65**, 119

DIETHYL N-BENZYLIDENEAMINOMETHYLPHOSPHONATE: PHOSPHONIC ACID,

    [[(PHENYLMETHYLENE)AMINO]METHYL]-, DIETHYL ESTER;

    (50917-73-2), **65**, 119

Diethyl 2-chloro-2-cyclopropylethene-1,1-dicarboxylate, **66**, 173, 174

N,N-DIETHYL-(E)-CITRONELLALENAMINE, (R)-(-)-: 1,6-OCTADIEN-1-AMINE,

    N,N-DIETHYL-3,7-DIMETHYL-, [R-(E)]-; (67392-56-7), **67**, 33

Diethyl cyclopropylcarbonylmalonate; (7394-16-3), **66**, 175, 179

Diethyl cyclopropylmalonate, **66**, 173

Diethyl dioxosuccinate; (59743-08-7), **66**, 144, 149

N,N-DIETHYLGERANYLAMINE; 2,6-OCTADIEN-1-AMINE, N,N-DIETHYL-3,7-

    DIMETHYL-, (E)-; (40267-53-6), **67**, 33, 44

Diethyl isocyanomethylphosphonate: Phosphonic acid, (isocyanomethyl)-, diethyl

    ester; (41003-94-5), **65**, 119

Diethyl malonate; (105-53-3), **66**, 175, 179

N,N-DIETHYLNERYLAMINE: 2,6-OCTADIEN-1-AMINE, N,N-DIETHYL-3,7-

    DIMETHYL-, (Z)-; (40137-00-6), **67**, 33, 48

Diethyl oxalate: Oxalic acid, diethyl ester; Ethanedioic acid, diethyl ester;
(95-92-1), **65**, 146

Diethyl phosphite; (762-04-9), **66**, 195, 197, 202

6-DIETHYLPHOSPHONOMETHYL-2,2-DIMETHYL-1,3-DIOXEN-4-ONE,
66, 194-196, 202

Diethyl phthalimidomethylphosphonate: Phosphonic acid, (phthalimidomethyl)-, diethyl ester; Phosphonic acid, [(1,3-dihydro-1,3-dioxo-2H-isoindol-2-yl)-methyl]-, diethyl ester; (33512-26-4), **65**, 119

DIHYDROJASMONE, **65**, 26

1,2-Dihydroxybenzene; (120-80-9), **66**, 180, 184

Dihydroxytartaric acid disodium salt hydrate; (866-17-1), **66**, 144, 146, 149

Diiodomethane; (75-11-6), **67**, 176

Diisobutylaluminum hydride; (1191-15-7), **66**, 60, 62, 66, 186, 188, 193

Diisopropylamine; 2-Propanamine, N-(1-methylethyl)-; (108-18-9), **65**, 98;
66, 37, 39, 88, 194; **67**, 125

DIISOPROPYL (2S,3S)-2,3-O-ISOPROPYLIDENETARTRATE: 1,3-DIOXOLANE-4,5-DICARBOXYLIC ACID, 2,2-DIMETHYL-, BIS(1-METHYLETHYL) ESTER, (4R-TRANS)-; (81327-47-1), **65**, 230

Diketene; (674-82-8), **66**, 109, 111, 114

Dimenthyl (1S,2S)-cyclopropane-1,2-dicarboxylate, (-)-: 1,2-Cyclopropane dicarboxylic acid, bis[5-methyl-2-(1-methylethyl)cyclohexyl] ester, [1S-[1S*,2S*,5R*)], 2β,5α]]-; (96149-01-8), **67**, 76

DIMENTHYL SUCCINATE, (-)-: BUTANEDIOIC ACID, BIS[5-METHYL-2-(1-METHYLETHYL)-CYCLOHEXYL]ESTER, [1R-[1α(1R*,2S*,5R*), 2β, 5α)]-; (34212-59-4), **67**, 76

1,1-Dimethoxyethylene: Ethene, 1,1-dimethoxy-; (922-69-0), **65**, 98

6,7-Dimethoxy-1,2,3,4-tetrahydro-2-[(1-tert-butoxy-3-methyl)-2-butylimino-
  methyl]isoquinoline: Isoquinoline 2-[[[1-[{1,1-dimethylethoxy)methyl]-2-
  methylpropyl]imino]methyl]-1,2,3,4-tetrahydro-6,7-dimethoxy-, (S)-;
  (90482-03-4), **67**, 60

6,7-Dimethoxy-1,2,3,4-tetrahydroisoquinoline hydrochloride: Isoquinoline,
  1,2,3,4-tetrahydro-6,7-dimethoxy-, hydrochloride; (2328-12-3), **67**, 60

N,N-Dimethylacetamide: Acetamide, N,N-dimethyl-; (127-19-5), **67**, 98

3,3-Dimethylallyl bromide; (870-63-3), **66**, 76, 78, 85

4-(N,N-Dimethylamino)pyridine: Pyridine, 4-(dimethylamino)-;
  4-Pyridinamine, N,N-dimethyl-; (1122-58-3), **65**, 12

N,N-DIMETHYL-N'-(1-tert-BUTOXY-3-METHYL-2-BUTYL)FORMAMIDINE,
  (S)-: METHANIMIDAMIDE, N'-[1-[(1,1-DIMETHYLETHOXY)METHYL]-2-
  METHYLPROPYL]-N,N-DIMETHYL-, (S)-; (90482-06-7), **67**, 52, 60

N,N-Dimethylchloromethylenammonium chloride; (3724-43-4), **66**, 121, 122, 124, 126

4,4-DIMETHYL-2-CYCLOPENTEN-1-ONE: 2-CYCLOPENTEN-1-ONE,
  4,4-DIMETHYL-; (22748-16-9), **67**, 121, 205

Dimethyl diazomethylphosphonate: Phosphonic acid, (diazomethyl)-, dimethyl ester:
  (27491-70-9), **65**, 119

2,2-Dimethyl-1,3-dioxane-4,6-dione, **67**, 170

3,3-DIMETHYL-1,5-DIPHENYLPENTANE-1,5-DIONE: 1,5-PENTANEDIONE,
  3,3-DIMETHYL-1,5-DIPHENYL-; (42052-44-8), **65**, 12

1,2-DIMETHYLENECYCLOHEXANE: CYCLOHEXANE, 1,2-DIMETHYLENE;
  CYCLOHEXANE, 1,2-BIS(METHYLENE)-; (2819-48-9), **65**, 90

N,N-Dimethylformamide, **66**, 121, 123, 195, 197

N,N-Dimethylformamide dimethyl acetal: Trimethylamine, 1,1-dimethoxy-;
  Methanamine, 1,1-dimethoxy-N,N-dimethyl-; (4637-24-5), **67**, 52

Dimethyl (E)-2-hexenedioate; (70353-99-0), **66**, 52, 53, 59

N,O-Dimethylhydroxylamine hydrochloride: Methylamine, N-methoxy-, hydrochloride; Methanamine, N-methoxy-, hydrochloride; (6638-79-5), **67**, 69

N,N-Dimethylisobutyramide, **66**, 117, 118

Dimethyl (2S,3S)-2,3-0-isopropylidenetartrate: 1,3-Dioxolane-4,5-dicarboxylic acid, 2,2-dimethyl-, dimethyl ester, (4R-trans)- or (4S-trans)-; (37031-29-1) or (37031-30-4), **65**, 230

Dimethyl malonate; (108-59-8), **66**, 76, 78, 85

Dimethyl methoxymethylenemalonate: Malonic acid, (methoxymethylene)-, dimethyl ester; Propanedioic acid, (methoxymethylene)-, dimethyl ester; (22398-14-7), **65**, 98

1,3-DIMETHYL-3-METHOXY-4-PHENYLAZETIDINONE: 2-AZETIDINONE, 3-METHOXY-1,3-DIMETHYL-4-PHENYL-; (82918-98-7), **65**, 140

1,3-Dimethyl-5-oxobicyclo[2.2.2]octane-2-carboxylic acid, **66**, 38

2,2-Dimethyl-4-oxopentanal: Pentanal, 2,2-dimethyl-4-oxo-; (61031-76-3), **67**, 121

N,N'-Dimethylpropyleneurea (DMPU); (7226-23-5), **66**, 45, 91, 94

Dimethyl sulfate; (77-78-1), **67**, 13

Dimethyl sulfide, **66**, 211

Dimethyl sulfoxide, **66**, 15, 17

1,9-DIMETHYL-8-(TRIMETHYLSILYL)BICYCLO[4.3.0]NON-8-EN-2-ONE, **66**, 8

DINAPHTHO[2,1-D:1'2'-f][1,3,2]DIOXAPHOSPHEPIN, 4-HYDROXY-4-OXIDE, R(-)-, **67**, 1

DINAPHTHO[2,1-d:1'2'-f][1,3,2]DIOXAPHOSPHEPIN, 4-HYDROXY-4-OXIDE, S(+)-, **67**, 1

Dinaphtho[2,1-d:1'2'-f]dioxaphosphepin, 4-methoxy 4-oxide, **67**, 13

1,3-Dioxane, 2-(bromomethyl)-2-(chloromethyl)-, **65**, 32

1,4-Dioxane, 2,3-dichloro-, trans-, **65**, 68

1,4-Dioxane, 2,3-bis(1,1-dimethylethoxy)-, trans-, **65**, 68

1,4-Dioxane, 2,3-bis(1,1-dimethylethoxy)-, cis-, , **65,** 68

1,3-Dioxane-4,6-dione, 2,2-dimethyl-, **67**, 170

1,3-Dioxane-4,6-dione, 5-(hexahydro-2H-azepin-2-ylidene)-2,2-dimethyl-
   67, 170

6,10-Dioxaspiro[4.5]dec-3-ene-1,1-dicarbonitrile, 2-phenyl-, **65**, 32

4,8-DIOXASPIRO[2.5]OCT-1-ENE, **65**, 32

p-Dioxino[2,3,-b]-p-dioxin, hexahydro, **65**, 68

[1,4]-Dioxino[2,3-b]-1,4-dioxin, hexahydro, **65**, 68

1,3-Dioxolane-4,5-dicarboxylic acid, 2,2-dimethyl-, dimethyl ester,
   (4R-trans)- or (4S-trans)-, **65**, 230

1,3-DIOXOLANE-4,5-DICARBOXYLIC ACID, 2,2-DIMETHYL-, BIS(1-METHYLETHYL)
   ESTER, (4R-TRANS)-, **65**, 230

2,6-DIOXO-1-PHENYL-4-BENZYL-1,4-DIAZABICYCLO[3.3.0]OCTANE:
   PYRROLO[3,4-c]PYRROLE-1,3(2H,3aH)-DIONE, TETRAHYDRO-
   2-PHENYL-5-(PHENYLMETHYL)-, cis-; (87813-00-1), **67**, 133

Diphenylmethylchlorosilane: Silane, chloromethyldiphenyl-; (144-79-6),
   67, 125

α-DIPHENYLMETHYLSILYLATION OF ESTER ENOLATES, **67**, 125

Diphenylphosphinyl chloride: Phosphinic chloride, diphenyl-; (1499-21-4),
   67, 20.

2,7-DISILAOCTA-3,5-DIYNE, 2,2,7,7-TETRAMETHYL-, **65**, 52

Disodium dihydrogen pyrophosphate; (7758-16-9), **66**, 214, 219

1,3-Dithiane; (505-23-7), **67**, 205

3,7-Dithianonane-1,9-diol: Ethanol, 2,2'-(trimethylenedithiol)di-;
   Ethanol, 2,2'-[1,3-propanediylbis(thio)]bis-;
   (16260-48-3), **65**, 150

3,7-Dithianonane-1,9-dithiol: Ethanethiol, 2,2'-(trimethylenedithio)di-;
Ethanethiol, 2,2'-[1,3-propanediylbis(thio)]bis-; (25676-62-4), **65**, 150

1-(1,3-Dithian-2-yl)-2-cyclohexen-1-ol: 2-Cyclohexen-1-ol, 1-(1,3-dithian-2-yl)-; (53178-46-4), **67**, 205

DMAP, **65**, 12

Dowex AG 50W-XB cation exchange resin, **66**, 212, 214, 215

Enamines, α–chloro, **66**, 119

Enamines, as Diels-Alder dienophiles, **66**, 147

β-Enamino esters, **67**, 170

Enolate chlorination, **66**, 194

Epoxide formation, from chlorohydrin, **66**, 160

Ester enolate alkylation, **66**, 87

Ester hydrolysis, **66**, 38, 87, 89

1,2-Ethanediamine, compound with lithium acetylide, **67**, 86

Ethane, hexachloro- (67-72-1), **66**, 202

Ethane, 1, 1',1"-[methylidynetris(oxy)]tris-, **66**, 146

Ethane, 1,1'-oxybis-, compd. with trifluoroborane (1:1), **65**, 17

Ethanedioic acid, diethyl ester, **65**, 146

Ethanethiol, 2,2'-[1,3-propanediylbis(thio)]bis-, **65**, 150

Ethanethiol, 2,2'-(trimethylenedithio)di-, **65**, 150

Ethanol, **66**, 175

Ethanol, 2-chloro-, **65**, 150

Ethanol, 2,2'-[1,3-propanediylbis(thio)]bis-, **65**, 150

Ethanol, 2,2'-(trimethylenedithio)di-, **65**, 150

Ethanone, 1-phenyl-, **65**, 6, 119

Ethene, 1,1-dimethoxy-, **65**, 98

1-(1-Ethoxyethoxy)-1,2-propadiene, metallation, **66**, 18

1-Ethoxy-1-trimethylsiloxycyclopropane: Silane, [(1-ethoxycyclopropyl)oxy]-, trimethyl-, **67**, 210

Ethyl 4-bromobutyrate: Butanoic acid, 4-bromo-, ethyl ester; (2969-81-5), **67**, 86

Ethyl α–(bromomethyl)acrylate; (17435-72-2), **66**, 222, 224

Ethyl chloroformate; (541-41-3), **66**, 133, 135, 141; **67**, 86

Ethyl cyanoformate; (623-49-4), **66**, 143, 149

ETHYL 4-CYCLOHEXYL-4-OXOBUTANOATE: CYCLOHEXANEBUTANOIC ACID, γ–OXO-, ETHYL ESTER; (54966-52-8), **65**, 17

ETHYL CYCLOPROPYLPROPIOLATE, **66**, 173, 174

ETHYL (E,Z)-2,4-DECADIENOATE; (3025-30-7), **66**, 22, 23, 25, 28

Ethyl 3,4-decadienoate; (36186-28-4), **66**, 22, 28

Ethyl decanoate: Decanoic acid, ethyl ester; (110-38-3), **67**, 125

Ethyl 2-(diphenylmethylsilyl)decanoate: Decanoic acid, 2-(methyldiphenylsilyl)-, ethy ester; (89638-16-4), **67**, 125

Ethylenediaminetetraacetic acid, tetrasodium salt: Glycine, N,N'-1,2-ethanediylbis[N-(carboxymethyl)]-, tetrasodium salt, trihydrate; (67401-50-7), **65**, 166

Ethyl ether, compd. with boron fluoride (BF$_3$) (1:1), **65**, 17

Ethyl formate: Formic acid, ethyl ester; (109-94-4), **67**, 52

ETHYL α-(HEXAHYDROAZEPINYLIDENE-2)ACETATE: ACETIC ACID, (HEXAHYDRO-2H-AZEPIN-2-YLIDENE)-, ETHYL ESTER, (Z)-; (70912-51-5), **67**, 170

ETHYL α-(HYDROXYMETHYL)ACRYLATE; (10029-04-6), **66**, 220, 222, 224

ETHYL 4-IODOBUTYRATE: BUTANOIC ACID, 4-IODO-, ETHYL ESTER; (7425-53-8) **67**, 98

Ethyl N-(2-methyl-3-nitrophenyl)formimidate, **65**, 146

ETHYL (E)-4-(4-NITROPHENYL)-4-OXO-2-BUTENOATE: 2-BUTENOIC ACID, 4-(4-NITROPHENYL)-4-OXO-, ETHYL ESTER, **67**, 86

Ethyl oxalamidrazonate; (53085-26-0), **66**, 143, 146, 149

ETHYL 5-OXO-6-METHYL-6-HEPTENOATE, **67**, 98

Ethyl propiolate; (623-47-2), **66**, 29, 31, 36

Ethyl thioamidooxalate; (16982-21-1), **66**, 143, 145, 149

Ethyne, **65**, 61

ETHYNYL p-TOLYL SULFONE: BENZENE, 1-(ETHYNYLSULFONYL)-4-METHYL-; (13894-21-8), **67**, 149

Finkelstein reaction, **66**, 87

Flash chromatography, **66**, 135, 196

Florisil, **66**, 197

Formaldehyde, **66**, 220

Formamide, N-[1-[(1,1-dimethylethoxy)methyl]-2-methylpropyl]-, (S)-, **67**, 52

Formamidines, N,N-dimethyl-N'-alkyl-, **67**, 52

Formic acid, chloro-, ethyl ester; (541-41-3), **66**, 141; **67**, 86

Formic acid, chloro-, methyl ester, **65**, 47

Formic acid, cyano-, ethyl ester; (623-49-4), **66**, 149

Formic acid, ethyl ester, **67**, 52

N-Formyl-O-tert-butylvalinol, (S)-: Formamide, N-[1-[(1,1-dimethylethoxy)-methyl]-2-methylpropyl]-, (S)-; (90482-04-5), **67**, 52

Geraniol; (106-24-1), **66**, 211, 214, 219

Geranyl chloride; (5389-87-7), **66**, 211, 212, 214, 219

GERANYL DIPHOSPHATE, TRISAMMONIUM SALT, **66**, 211-213

Glucopyranoside, methyl, α–D-. **65**, 243

Glucopyranoside, methyl 4,6-O-benzylidene-, α–D-, **65**, 243

GLUCOPYRANOSIDE, METHYL 6-BROMO-6-DEOXY, 4-BENZOATE, α–D-, **65**, 243

Glucopyranoside, methyl 4,6-O-(phenylmethylene)-, α–D-, **65**, 243

Glucopyranosyl bromide, 2,3,4,6-tetraacetate, α–D-, **65**, 236

Glucopyranosyl bromide tetraacetate, α–D-, **65**, 236

GLYCERO-D-IDO-NONONONITRILE, 4,8-ANHYDRO-2,3-DIDEOXY-, 5,6,7,9-TETRAACETATE, α–D-, **65**, 236

Glycine, N,N'-I,2-ethanediylbis[N-(carboxymethyl)]-, tetrasodium salt, trihydrate, **65**, 166

Grignard reagents, reaction with acyl chlorides, **66**, 116

Grob Fragmentation, **66**, 173

Halide exchange reaction, **66**, 87

2,5-Heptadien-4-ol, 3,4,5-trimethyl-, **65**, 42

Heptanal; (111-71-7), **65**, 26

3-HEPTANONE, 4-METHYL-, (S)-, **65**, 183

5-Heptynoic acid, 7-hydroxy-, methyl ester, **67**, 193

Hexacarbonyl(1-methyl-2-propynylium)dicobalt tetrafluoroborate: Cobalt(1+), hexacarbonyl[μ-[(2,3-η:2,3-η)-1-methyl-2-propynylium]]di-, (Co-Co), tetrafluoroborate(1-); (62866-98-2), **67**, 141

HEXACARBONYL(PROPARGYLIUM)DICOBALT SALTS, **67**, 141

Hexachloroethane; (67-72-1), **66**, 195, 197, 202

(5Z,7E)-5,7-HEXADECADIENE, **66**, 60, 61, 63

HEXAHYDRO-4,4,7-TRIMETHYL-4H-1,3-BENZOXATHIIN: 4H-1,3-BENZOXATHIIN, HEXAHYDRO-4,4,7-TRIMETHYL-; (59324-06-0), **65**, 215

Hexamethylphosphoric triamide (HMPA); (680-31-9), **66**, 44, 45, 51, 88, 94

Hexanedioic acid, monomethyl ester; (627-91-8), **66**, 120

2,4-Hexenedioic acid, monomethyl ester (Z,Z)-; (61186-96-7), **66**, 184

1-Hexene, 1-iodo-, (E); (16644-98-7), **66**, 66

1-Hexene, 1-iodo-, (Z); (16538-47-9), **66**, 66

(E)-1-Hexenyldiisobutylalane; (20259-40-9), **66**, 66

(E)-1-Hexenyl iodide; (16644-98-7), **66**, 63, 66

(Z)-1-Hexenyl iodide; (16538-47-9), **66**, 61, 62, 66

1-Hexyne; (693-02-7), **66**, 66

    reaction with diisobutylaluminum hydride, **66**, 63

Higher order cuprates, **66**, 57

HOFMANN REARRANGEMENT UNDER MILDLY ACIDIC CONDITIONS, **66**, 132

Homoenolate, copper, **66**, 47

Homoenolate, titanium, **66**, 47

HOMOENOLATE, ZINC, **66**, 43

Horner-Wadsworth-Emmons reaction, **66**, 220

Hydrazine, **66**, 143; **67**, 60, 187

Hydrazine, tert-butyl, monohydrochloride, **65**, 166

Hydrazine, (1,1-dimethylethyl)-, monohydrochloride, **65**, 166

Hydroalumination of alkynes, **66**, 64

Hydroboration of alkynes, **66**, 64

Hydrogen chloride, **66**, 144, 146

Hydrogen peroxide; (7722-84-1), **65**, 166

Hydrogen sulfide, **66**, 143, 145

Hydroquinone: 1,4-Benzenediol; (123-31-9), **67**, 98

    as diene stabilizer, **66**, 63

3-HYDROXY-1-CYCLOHEXENE-1-CARBOXALDEHYDE: 1-CYCLOHEXENE-
    1-CARBOXALDEHYDE, 3-HYDROXY-; (67252-14-6), **67**, 205

Hydroxylamine, N-phenyl-, **67**, 187

α-Hydroxylation of ketones, **66**, 138

1-HYDROXYMETHYL-4-(1-METHYLCYCLOPROPYL)-1-CYCLOHEXENE:
1-CYCLOHEXENE-1-METHANOL, 4-(1-METHYLCYCLOPROPYL)-;
(536-59-4), **67**, 176

3-HYDROXY-3-METHYL-1-PHENYL-1-BUTANONE: 1-BUTANONE,
3-HYDROXY-3-METHYL-1-PHENYL-; (43108-74-3), **65**, 6, 12

N-Hydroxymethylphthalimide: Phthalimide, N-(hydroxymethyl)-;
1H-Isoindole-1,3-(2H)-dione, 2-(hydroxymethyl)-; (118-29-6), **65**, 119

(1-Hydroxy-2-propenyl)trimethylsilane, **66**, 14, 15

Hydrozirconation of alkynes, **66**, 64

Hypervalent iodine compounds, **66**, 138

Imines, reduction by diisobutylaluminum hydride, **66**, 189

5H-Indene-5-one, 1,2,3,3a,4,6-hexahydro-, **67**, 163

INDOLE, 4-NITRO-, **65**, 146

INTRAMOLECULAR ACYLATION OF ALKYLSILANES, **66**, 87

INVERSE ELECTRON DEMAND DIELS-ALDER, **65**, 98

Iodine, phenylbis(trifluoroacetato-O); (2712-78-9), **66**, 141

Iodobenzene diacetate, **66**, 136

Iodobenzene dichloride, **66**, 137

6-Iodo-3,4-dimethoxybenzaldehyde cyclohexylimine: Cyclohexanamine,
N-[(2-iodo-4,5-dimethoxyphenyl)methylene]-; (61599-78-8), **65**, 108

o-Iodotoluene; (615-37-2), **66**, 67, 68, 74

1-Iodo-3-trimethylsilylpropane; (18135-48-3), **66**, 87, 88, 91, 94

Ion exchange chromatography, **66**, 212, 214, 215

Iron pentacarbonyl, **66**, 99

Isobutene; (115-11-7), **67**, 52

Isobutyl chloroformate, **66**, 135

(R)-Isobutyloxirane, **66**, 165

Isoindole-1,3-(2H)-dione, 2-(bromomethyl)-, 1H-, **65**, 119

Isoindole-1,3-(2H)-dione, 2-(hydroxymethyl)-, 1H-, **65**, 119

(S)-Isoleucine, **66**, 153

Isomenthol, (+)-: Cyclohexanol, 5-methyl-2-(1-methylethyl)-, [1S-(1α,2β,5β)];
  (23283-97-8), **65**, 81

Isomenthone, (+)-: Cyclohexanone, 5-methyl-2-(1-methylethyl)-, (2R-cis)-;
  (1196-31-2), **65**, 81

ISOPRENE; (78-79-5), **67**, 48

Isopropyl alcohol, titanium (4+) salts, **65**, 230; **67**, 180

Isopropylideneacetophenone: 2-Buten-1-one, 3-methyl-1-phenyl-;
  (5650-07-7), **65**, 12

Isopropylidene α-(hexahydroazepinylidene-2)malonate: 1,3-Dioxane-4,6-
  dione, 5-(hexahydro-2H-azepin-2-ylidene)-2,2-dimethyl-; (70192-54-8),
  67, 170

(R)-Isopropyloxirane, **66**, 165

Isoquinoline, 1,2,3,4-tetrahydro-6,7-dimethoxy, hydrochloride, **67**, 60

Isoquinoline, 1,2,3,4-tetrahydro-6,7-dimethoxy-1-methyl-, (S)-, **67**, 60

Ketones, preparation from carboxylic acid, **66**, 119

β–Lactams, **65**, 140

Lactic acid, 2-methyl-, methyl ester; (2110-78-3), **66**, 114

Lead dioxide: Lead oxide; (1309-60-0), **65**, 166

Lead oxide, **65**, 166

(S)-Leucine, **66**, 153

Leucine, N-carboxy-, N-tert-butyl ester, L-, **67**, 69

L-Leucine, N-[(1,1-dimethylethoxy)carbonyl]-, **67**, 69

Lipshutz reagents, **66**, 57

Lithiobutadiyne, **65**, 52

Lithium, **66**, 127, 128; **67**, 193

Lithium acetate dihydrate: Acetic acid, lithium salt, dihydrate; (6108-17-4), **67**, 105

Lithium acetylide, ethylenediamine complex: 1,2-Ethanediamine, compound with lithium acetylide (Li(C_2H)); (6867-30-7), **67**, 86

Lithium aluminum hydride, **66**, 160; **67**, 69

Lithium, butyl-, **65**, 98, 108, 119; **66**, 14, 16, 37, 39, 88, 194, 196; **67**, 44, 48

Lithium, tert-butyl-, **66**, 15, 16, 18, 67, 69; **67**, 60, 210

Lithium diisopropylamide, **65**, 98; **66**, 37, 88, 194; **67**, 125

Lithium fluoride, **67**, 133

Lithium, methyl-, **65**, 47, 140; **67**, 86, 125

Lithium tetrafluoroborate, **66**, 52, 54

Lithium 2,2,6,6-tetramethylpiperidide (LTMP), **67**, 76

Lithium tri(tert-butoxy)aluminum hydride, **66**, 122, 124

Lyophilization, **66**, 212-214

MACROCYCLIC SULFIDES, **65**, 150

Magnesium, **66**, 118, 175

Magnesium ethoxide, **66**, 175

Maleimide, N-phenyl-, **67**, 133

Malonic acid, cyclic isopropylidene ester, **67**, 170

Malonic acid, dimethyl ester; (108-59-8), **66**, 85

Malonic acid, (methoxymethylene)-, dimethyl ester, **65**, 98

Malonic ester alkylation, **66**, 75

Malononitrile, benzylidene-, **65**, 32

MELDRUM'S ACID: 2,2-DIMETHYL-1,3-DIOXANE-4,6-DIONE; MALONIC ACID, CYCLIC ISOPROPYLIDENE ESTER; 1,3-DIOXANE-4,6-DIONE, 2,2-DIMETHYL-; (2033-24-1), **67**, 170

p-Mentha-6,8-dien-2-one; (6485-40-1), **66**, 13

p-MENTH-4-(8)-EN-3-ONE, (R)-(+)-, **65**, 203, 215

Menthol, (-)-: Cyclohexanol, 5-methyl-2-(1-methylethyl)-, [1R-(1α,2β,5α)]–; (2216-51-5), **67**, 76

2-(1-Mercapto-1-methylethyl)-5-methylcyclohexanol: Cyclohexanol-2-(1-mercapto-1-methylethyl)-5-methyl-, [1R-(1α,2α,5α)]-; (79563-68-1); [1R-(1α,2β,5α)] (79563-59-0); [1S-(1α,2α,5β)]-; (79563-67-0), **65**, 215

Mercury, **66**, 96

Mercury(II) oxide, red; (21908-53-2), **67**, 205

METHACRYLOYL CHLORIDE: 2-PROPENOYL CHLORIDE, 2-METHYL-; (920-46-7), **67**, 98

Methanamine, 1,1-dimethoxy-N,N-dimethyl-, **67**, 52

Methanamine, N-methoxy-, hydrochloride, **67**, 69

Methanamine, N-(phenylmethylene)-, **65**, 140

Methane, dibromo-, **65**, 81

METHANESULFONYL BROMIDE, BROMO-, **65**, 90

Methanesulfonyl chloride; (124-63-0), **66**, 1, 3, 186, 187, 193

Methanethiol, **66**, 188

Methanimidamide, N'-[1-[(1,1-dimethylethoxy)methyl]-2-methylpropyl]-N,N-dimethyl-, (S)-, **67**, 52, 60

Methanol; (67-56-1), **66**, 85, 182, 184

N^{1}N^{2}Bis(methoxycarbonyl)sulfur diimide: Sulfur diimide, dicarboxy-, dimethyl ester; (16762-82-6), **65**, 159

6-Methoxy-7-methoxycarbonyl-1,2,3,4-tetrahydronaphthalene: 2-Naphthalenecarboxylic acid, 5,6,7,8-tetrahydro-3-methoxy-, methyl ester; (78112-34-2), **65**, 98

4-METHOXY-3-PENTEN-2-ONE: 3-PENTEN-2-ONE 4-METHOXY-; (2845-83-2), **67**, 202

1-Methoxy-3-(trimethylsiloxy)-1,3-butadiene: Silane, [(3-methoxy-1-methylene-2-propenyl)oxy]trimethyl-, (59414-23-2), **67**, 163

Methyl acrylate; (96-33-3), **66**, 54, 59

  dimerization by Pd(II), **66**, 52

Methylamine, N-benzylidene-, **65**, 140

Methylamine, N-methoxy-, hydrochloride, **67**, 69

METHYL 4-O-BENZOYL-6-BROMO-6-DEOXY-α–D-GLUCOPYRANOSIDE: GLUCOPYRANOSIDE, METHYL 6-BROMO-6-DEOXY, 4-BENZOATE, α–D-; (10368-81-7), **65**, 243

Methyl 4,6-O-benzylidene-α–D-glucopyranoside: Glucopyranoside, methyl 4,6-O-benzylidene-α–D-; α–D-glucopyranoside, methyl 4,6-O-(phenylmethylene)-; (3162-96-7), **65**, 243

N-Methylbenzylidenimine: Methylamine, N-benzylidene-; Methanamine, N-(phenylmethylene)-; (622-29-7), **65**, 140

Methyl 4-bromo-1-butanimidate hydrochloride, **67**, 193

2-Methyl-2-butene: 2-Butene, 2-methyl-; (513-35-9), **65**, 159

(3-Methyl-2-butenyl)propanedioic acid, dimethyl ester; (43219-18-7), **66**, 78, 85

(3-Methyl-2-butenyl)(2-propynyl)propanedioic acid, dimethyl ester, **66**, 76

O-METHYLCAPROLACTIM: 2H-AZEPINE, 3,4,5,6-TETRAHYDRO-7-
    METHOXY-; (2525-16-8); **67**, 170

Methyl carbamate: Carbamic acid, methyl ester; (598-55-0), **65**, 159

METHYL 4-CHLORO-2-BUTYNOATE: 2-BUTYNOIC ACID, 4-CHLORO-, METHYL
    ESTER; (41658-12-2), **65**, 47

Methyl chloroformate: Formic acid, chloro-, methyl ester;
    Carbonochloridic acid, methyl ester; (79-22-1), **65**, 47

Methyl (E)-crotonate; (623-43-8), **66**, 38, 39, 41, 42

1-Methylcyclohexene: Cyclohexene, 1-methyl-; (591-49-1), **65**, 90

3-Methyl-2-cyclohexen-1-one; (1193-18-6), **66**, 37, 39, 42

Methyl N,N-dichlorocarbamate: Carbamic acid, dichloro-, methyl ester;
    (16487-46-0), **65**, 159

Methyl 1,3-dimethyl-5-oxobicyclo[2.2.2]octane-2-carboxylate, **66**, 37

Methyl 1,1'-dinaphthyl-2,2'-diyl phosphate, (R)-(-)-: Dinaphtho[2,1-d:1'2'-f]di-
    oxaphosphepin, 4-methoxy-, 4-oxide, (R)-; (86334-02-3), **67**, 13

METHYLENATION OF CARBONYL COMPOUNDS, **65**, 81

3-Methylene-4-isopropyl-1,1-cyclopentanedicarboxylic acid,
    dimethyl ester, **66**, 78

3-METHYLENE-CIS-p-MENTHANE, (+)-: (CYCLOHEXANE,
    5-METHYL-1-METHYLENE-2-(1'-METHYLETHYL)-, R,R-), **65**, 81

Methyl α–D-glucopyranoside: Glucopyranoside, methyl, α–D-;
    α–D-glucopyranoside, methyl; (97-30-3), **65**, 243

4-METHYL-3-HEPTANONE, (S)-(+)-: 3-HEPTANONE, 4-METHYL-, (S)-;
    (51532-30-0), **65**, 183

4-Methyl-3-heptanone SAMP-hydrazone, (S)-(+)-: 1-Pyrrolidinamine,
    N-(1-ethyl-2-methylpentylidene)-2-(methoxymethyl)-,
    [S-[R*,R*-(Z)]-; (69943-24-4), **65**, 183

METHYL 7-HYDROXYHEPT-5-YNOATE: 5-HEPTYNOIC ACID, 7-HYDROXY-,
    METHYL ESTER; (50781-91-4), **67**, 193
Methyl 2-hydroxyisobutyrate; (2110-78-3), **66**, 110, 111, 114
Methyllithium: Lithium, methyl-; (917-54-4), **65**, 47, 140,; **66**, 97, 100; **67**, 86, 125
Methyllithium, low halide, **66**, 53, 55
Methylmagnesium bromide, **67**, 125
Methylmagnesium chloride, **66**, 1-3
[(Methyl)(methoxy)carbene]pentacarbonyl chromium(0): Chromium,
    pentacarbonyl(1-methoxyethylidene)-, (OC-6-21)-; (20540-69-6), **65**, 140
METHYL N-(2-METHYL-2-BUTENYL)CARBAMATE: CARBAMIC ACID, (2-METHYL-
    2-BUTENYL)-, METHYL ESTER; (86766-65-6), **65**, 159
5-Methyl-2-(1-methyl-1-phenylethyl)cyclohexanone, (2R,5R)-: Cyclohexanone,
    5-methyl-2-(1-methyl-1-phenylethyl)-, (2R-trans)-; (57707-92-3), **65**, 203
5-Methyl-2-(1-methyl-1-phenylethyl)cyclohexanone, (2S,5R)-: Cyclohexanone,
    5-methyl-2-(1-methyl-1-phenylethyl-, (2S-cis)-; (65337-06-6), **65**, 203
5-Methyl-2-(1-methyl-1-phenylethyl)cyclohexyl chloroacetate, (1R,2S,5R)-:
    Acetic acid, chloro-, 5-methyl-2-(1-methyl-1-phenylethyl)cyclohexyl ester,
    [1R-(1α,2β,5α)]-; (71804-27-8), **65**, 20
5-Methyl-2-[1-methyl-1-(phenylmethylthio)ethyl]cyclohexanone, cis- and trans-;
    65, 215
5-Methyl-2-(1-methyl-1-thioethyl)cyclohexanol, **65**, 215
N-Methylmorpholine, **66**, 133, 135
2-Methyl-3-nitroaniline: o-Toluidine, 3-nitro-; Benzeneamine,
    2-methyl-3-nitro-; (603-83-8), **65**, 146
2-METHYL-4'-NITROBIPHENYL, **66**, 67, 68
(R)-METHYLOXIRANE; (15448-47-2), **66**, 160, 161, 163, 164, 172
METHYL 6-OXODECANOATE; (61820-00-6), **66**, 116, 117, 120, 123

METHYL 2-OXO-5,6,7,8-TETRAHYDRO-2H-1-BENZOPYRAN-3-CARBOXYLATE:
2H-1-BENZOPYRAN-3-CARBOXYLIC ACID, 5,6,7,8-TETRAHYDRO-2-OXO-,
METHYL ESTER; (85531-80-2), **65**, 98

(E)-3-Methyl-3-penten-2-one, **66**, 11

(Z)-3-Methyl-3-penten-2-one, **66**, 11

3-METHYL-2-PENTYL-2-CYCLOPENTEN-1-ONE: 2-CYCLOPENTEN-1-ONE,
3-METHYL-2-PENTYL-; (1128-08-1), **65**, 26

2-METHYL-2-PHENYL-4-PENTENAL: 4-PENTENAL, 2-METHYL-2-PHENYL-;
(24401-39-6), **65**, 119

N-Methylpiperidine: 1-Methylpiperidine; (626-67-5), **67**, 69

2-Methylpropane-2-thiol; (75-66-1), **66**, 108, 109, 111, 114

2-(1-METHYL-2-PROPYNYL)CYCLOHEXANONE, **67**, 141

Methyl tiglate; (6622-76-0), **66**, 88, 91, 94

Methyltriisopropoxytitanium: Titanium, triisopropoxymethyl-; Titanium,
methyltris(2-propanolato)-, (T-4)-; (19006-13-8), **67**, 180

1-Methyl-1-(trimethylsilyl)allene, **66**, 1, 2, 4

2-METHYL-2-UNDECENE: 2-UNDECENE, 2-METHYL-; (56888-88-1),
67, 125

2-METHYL-2-VINYLCYCLOPENTANONE; (88729-76-4), **66**, 87, 88, 90, 91, 94

trans-3-METHYL-2-VINYLCYCLOHEXANONE, **66**, 95, 98

MICHAEL ADDITION, APROTIC, **66**, 37, 41

cis,cis-Monomethyl muconate; (61186-96-7), **66**, 180, 184

cis,cis-MONOMETHYL MUCONATE FROM 1,2-DIHYDROXYBENZENE, **66**, 180

Morpholine; (110-91-8), **67**, 222

Morpholine 4-chloro-, **67**, 222

MYRCENE: 1,6-OCTADIENE, 7-METHYL-3-METHYLENE-; (123-35-3),
67, 44

2-Naphthalenecarboxylic acid, 5,6,7,8-tetrahydro-3-methoxy-, methyl ester, **65**, 98

Nickel acetylacetonate: Nickel, bis(2,4-pentanedionato-O,O')-, (SP-4-1);
(3264-82-2), **67**, 170

Nickel-catalyzed aryl-aryl coupling, **66**, 70

m-Nitrobenzaldehyde: Benzaldehyde, 3-nitro-; (99-61-6), **67**, 180

p-Nitrobenzoyl chloride: Benzoyl chloride, 4-nitro-; (122-04-3), **67**, 86

4-NITROINDOLE: INDOLE, 4-NITRO-; (4769-97-5), **65**, 146

3'-NITRO-1-PHENYLETHANOL: BENZENEMETHANOL, α-METHYL-3-NITRO-;
(5400-78-2), **67**, 180

1-Nitroso-2-methoxymethylpyrrolidine, (S)-: Pyrrolidine, 2-(methoxymethyl)-
1-nitroso-, (S)-; (60096-50-6), **65**, 183

Nitrosonium tetrafluoroborate, **66**, 54

Nitroso-tert-octane: Pentane, 2,2,4-trimethyl-4-nitroso-; (31044-98-1), **65**, 166

1,6-Octadien-1-amine, N,N-diethyl-3,7-dimethyl, [R-(E)]-, **67**, 33

2,6-Octadien-1-amine, N,N-diethyl-3,7-dimethyl, (E)-, **67**, 33, 44

2,6-Octadien-1-amine, N,N-diethyl-3,7-dimethyl-, (Z)-, **67**, 33, 48

1,6-Octadiene, 7-methyl-3-methylene-, **67**, 44

6-Octenal, 3,7-dimethyl-, (R)-(+)-, **67**, 33

tert-Octylamine: 2-Pentanamine, 2,4,4-trimethyl-; (107-45-9), **65**, 166

N-tert-Octyl-O-tert-butylhydroxylamine: 2-Pentanamine, N-(1,1-dimethylethoxy)-
2,4,4-trimethyl-; (68295-32-9), **65**, 166

1-Octyn-3-ol; (818-72-4), **66**, 22, 23, 28

Organoaluminum compounds, reaction with imino carbocations, **66**, 189

Orthoester Claisen rearrangement, **66**, 22

Orthoformic acid, triethyl ester, **65**, 146

6-Oxabicyclo[3.1.0]hex-2-ene, **67**, 114

Oxalic acid, diethyl ester, **65**, 146

Oxalyl chloride; (79-37-8), **66**, 15, 17, 21, 89, 94, 117, 121, 123

1,3-OXATHIANE, **65**, 215

Oxaziridine, 3-phenyl-2-(phenylsulfonyl)-; (63160-13-4), **66**, 203, 210

OXIDATIVE CLEAVAGE OF AROMATIC RINGS; **66**, 180

Oximes, mesylation, **66**, 185

Oxirane, methyl; (15448-47-2), **66**, 172

4-Oxo-1-(benzenesulfonyl)-cis-bicyclo[4.3.0]non-2-ene, **67**, 163

6-OXODECANAL; (63049-53-6), **66**, 121, 122, 126

6-Oxodecanoic acid; (4144-60-9), **66**, 122, 123, 126

Oxonium, trimethyl-, tetrafluoroborate (1-), **65**, 140

(1-OXO-2-PROPENYL)TRIMETHYLSILANE, **66**, 14-16, 18

Ozone (10028-15-6), **65**, 183

Palladium acetate: Acetic acid, palladium(2+) salt; (3375-31-3), **67**, 105

Palladium, benzylchlorobis(triphenylphosphine)-, trans-, **67**, 86

Palladium-catalyzed aryl-aryl coupling, **66**, 70

PALLADIUM-CATALYZED ALLYLIC AMINATION, **67**, 105

PALLADIUM-CATALYZED CHLOROACETOXYLATION, **67**, 105

PALLADIUM-CATALYZED COUPLING OF ACID CHLORIDES WITH
    ORGANOTIN REAGENTS, **67**, 86

PALLADIUM-CATALYZED COUPLING OF ARYL HALIDES, **66**, 67

PALLADIUM-CATALYZED syn-ADDITION OF CARBOXYLIC ACIDS,
    67, 114

Palladium(II) chloride; (7647-10-1), **67**, 121

Palladium, chloro(phenylmethyl)bis(triphenylphosphine)-, **67**, 86

Palladium sponge, **66**, 54

Palladium, tetrakis(acetonitrile)-, tetrafluoroborate, **66**, 52

Palladium, tetrakis(triphenylphosphine)-, **67**, 86, 98, 105, 114

Palladium, tetrakis(triphenylphosphine)-; (14221-01-3),
    66, 61, 62, 66, 68, 69, 74

Paraformaldehyde: Poly(oxymethylene); (9002-81-7), **65**, 215, **66**, 220, 221

1,2,3,4,5-PENTAMETHYLCYCLOPENTADIENE: 1,3-CYCLOPENTADIENE,
    1,2,3,4,5-PENTAMETHYL-; (4045-44-7), **65**, 42

Pentanal, 2,2-dimethyl-4-oxo-, **67**, 121

2-Pentanamine, N-(1,1-dimethylethoxy)-N-(1,1-dimethylethyl)-
    2,4,4-trimethyl-, **65**, 166

2-Pentanamine, N-(1,1-dimethylethoxy)-2,4,4-trimethyl-, **65**, 166

2-PENTANAMINE, N-(1,1-DIMETHYLETHYL)-2,4,4-TRIMETHYL-, **65**, 166

2-Pentanamine, 2,4,4-trimethyl-, **65**, 166

1,5-PENTANEDIONE, 3,3-DIMETHYL-1,5-DIPHENYL-, **65**, 12

Pentane, 2,2,4-trimethyl-4-nitroso-, **65**, 166

3-Pentanone SAMP-hydrazone: 1-Pyrrolidinamine, N-(1-ethylpropylidene)-2-
    (methoxymethyl)-, (S)-; (59983-36-7), **65**, 183

4-PENTENAL, 2-METHYL-2-PHENYL-, **65**, 119

3-Penten-2-one, 4-methoxy-, **67**, 202

Perchloric acid, silver(1+) salt, monohydrate, **67**, 33

PERRILLYL ALCOHOL, (S)-(-)-: 1-CYCLOHEXENE-1-METHANOL, 4-
    (1-METHYLETHENYL)-; (536-59-4), **67**, 176

Phenylbenzeneselenosulfonate: Benzenesulfonoselenoic acid, Se-phenyl
    ester; (60805-71-2), **67**, 157

N-Phenylhydroxylamine: Hydroxylamine, N-phenyl-; Benzeneamine,
    N-hydroxy-; (100-65-2), **67**, 187

N-Phenylmaleimide: Maleimide, N-phenyl-; 1H-Pyrrole-2,5-dione, 1-phenyl-;
    (941-69-5), **67**, 133

8-PHENYLMENTHOL, (-)-: CYCLOHEXANOL, 5-METHYL-2-(1-METHYL-
    1-PHENYLETHYL)-, [1R-(1α,2β,5α)]-; (65253-04-5), **65**, 203

3-PHENYL-4-PENTENAL, **66**, 29, 31, 36

2-Phenyl-N-(phenylmethylene)-1-propen-1-amine: 1-Propen-1-amine,
    2-Phenyl-N-(phenylmethylene)-; (64244-34-4), **65**, 119

(E)-3-[(E)-3-Phenyl-2-propenoxy]acrylic acid; (88083-18-5), **66**, 30, 31, 36

Phenylseleninic acid: Benzeneseleninic acid; (6996-92-5), **67**, 157

(±)-trans-2-(PHENYLSULFONYL)-3-PHENYLOXAZIRIDINE, **66**, 203, 204

1-Phenyl-1-trimethylsiloxyethylene: Silane, trimethyl [(1-phenylvinyl)oxy]-;
    Silane, trimethyl[(1-phenylethenyl)oxy]-; (13735-81-4), **65**, 6

Phosphine, [1,1'-binaphthalene]-2,2'-diylbis[diphenyl-, (R)- or (S)-], **67**, 20, 33

Phosphine oxide, [1,1'-binaphthalene]-2,2'-diylbis[diphenyl-, (±)-, (S)-, and
    (R)-], **67**, 20

Phosphinic chloride, diphenyl-, **67**, 20

Phosphonate synthesis, **66**, 194

Phosphonic acid, (aminomethyl), diethyl ester, **65**, 119

Phosphonic acid, (diazomethyl)-, dimethyl ester, **65**, 119

Phosphonic acid, diethyl ester; (762-04-9), **66**, 202

Phosphonic acid, [(1,3-dihydro-1,3-dioxo-2H-isoindol-2-yl)methyl]-,
    diethyl ester, **65**, 119

Phosphonic acid, [(2,2-dimethyl-4-oxo-4H-1,3-dioxin-6-yl)methyl]-, diethyl
    ester; (81956-28-7), **66**, 22

Phosphonic acid, (isocyanomethyl)-, diethyl ester, **65**, 119

PHOSPHONIC ACID [[(PHENYLMETHYLENE)AMINO]METHYL]-, DIETHYL ESTER,
    65, 119

Phosphonic acid, (phthalimidomethyl)-, diethyl ester, **65**, 119

Phosphoric triamide, hexamethyl-; (680-31-9), **66**, 51, 94

Phosphorodichloridic acid, 2-chloroethyl ester, **65**, 68

Phosphorous acid, triethyl ester, **65**, 108. 119

Phosphorus oxychloride, **66**, 173, 176; **67**, 1

Phosphorus tribromide; (7789-60-8), **67**, 210

Phthalimide, N-(bromoethyl)-, **65**, 119

Phthalimide, N-(hydroxymethyl)-, **65**, 119

PINENE, (-)-α–: 2-PINENE, (1S,5S)-(-)-; BICYCLO[3.1.1]HEPT-2-ENE, 2,6,6-TRIMETHYL-, (1S)-; (7785-26-4), **65**, 224

PINENE, (-)-β–: BICYCLO[3.1.1]HEPTANE, 6,6-DIMETHYL-2-METHYLENE-, (1S)-; (18172-67-3), **65**, 224

Piperidine, 1-methyl, **67**, 69

Piperidine, 2,2,6,6-tetramethyl-, **67**, 76

Poly(oxymethylene), **65**, 215

Potassium 3-aminopropylamide (KAPA), **65**, 224

Potassium tert-butoxide, **66**, 127, 128, 195; **67**, 125

Potassium hydride; (7693-26-7), **65**, 224

Potassium hydroxide, **66**, 89

Prenyl bromide, **66**, 76

Proline, D-; (344-25-2), **65**, 173

Proline, L-; (147-85-3), **65**, 173

2-Propanamine, N-(1-methylethyl)-, **65**, 98

Propane, 1-bromo-3-chloro-2,2-dimethoxy-, **65**, 32

Propane, 2,2'-[(1-chloro-1,2-ethenediyl)bis(oxy)]bis[2-methyl-, (E)-, **65**, 68

Propane, 2,2'-[(1,2-dichloro-1,2-ethanediyl)]bis(oxy)bis[2-methyl-, (R*,R*)-(±)-, **65**, 68

Propane, 2,2'-[(1,2-dichloro-1,2-ethanediyl)bis(oxy)]bis[2-methyl-,
   (R*,S*)-, **65**, 68

PROPANE, 2,2'-[1,2-ETHYNEDIYLBIS(OXY)]BIS[2-METHYL-, **65**, 68

1,3-Propanediamine; (109-76-2), **65**, 224; **66**, 131

Propanedinitrile, (phenylmethylene)-, **65**, 32

Propanedioic acid, (cyclopropylcarbonyl)-, diethyl ester;
   (7394-16-3), **66**, 179

Propanedioic acid, diethyl ester; (105-53-3), **66**, 179

Propanedioic acid, (methoxymethylene)-, dimethyl ester, **65**, 98

Propanedioic acid, (3-methyl-2-butenyl)-, dimethyl ester;
   (43219-18-7), **66**, 85

1,3-Propanediol; (504-63-2), **65**, 32

1,3-Propanedithiol; (109-80-8), **65**, 150

2-Propanethiol, 2-methyl-; (75-66-1), **66**, 114

1-Propanol, 2-chloro-, (S)-(+)-; (19210-21-0), **66**, 172

Propanoic acid, 2-chloro-, (S)-; (29617-66-1), **66**, 159, 172

Propanoic acid, 2-hydroxy-2-methyl-, methyl ester; (2110-78-3), **66**, 114

2-Propanol, titanium (4+) salt, **65**, 230; **67**, 180

2-Propanone, 1-bromo-3-chloro-, dimethyl acetal, **65**, 32

Propargyl alcohol: 2-Propyn-1-ol; (107-19-7), **67**, 193

Propargyl bromide; (160-96-7), **66**, 77, 79, 86

PROPARGYL CHLORIDE: PROPYNE, 3-CHLORO-; 1-PROPYNE, 3-CHLORO-;
   (624-65-7), **65**, 47

1-Propen-1-amine, 1-chloro-N,N,2-trimethyl-; (26189-59-3), **66**, 120

1-Propen-1-amine, 2-phenyl-N-(phenylmethylene)-, **65**, 119

2-Propenenitrile, **65**, 236

2-Propenoic acid, 2-(bromomethyl)-, ethyl ester; (17435-72-2), **66**, 224

2-Propenoic acid, 2-(hydroxymethyl)-, ethyl ester; (10029-04-6), **66**, 224

2-Propen-1-ol; (107-18-6), **66**, 21

2-Propen-1-ol, 1-(trimethylsilyl)-; (95061-68-0), **66**, 21

2-Propenoyl chloride, 2-methyl-, **67**, 98

2-(1-PROPENYL)CYCLOBUTANONE, (E)-: CYCLOBUTANONE, 2-(1-
PROPENYL)-, (E)-; (63049-06-9), **67**, 210

Propionitrile, 2,2'-azobis[2-methyl-, **67**, 86

2-Propyl-1-azacycloheptane; (85028-29-1), **66**, 186-188, 193

2-PROPYL-1-AZACYCLOHEPTANE FROM CYCLOHEXANONE OXIME, **66**, 185

Propylene oxide, (R)-(+)-; (15448-47-2), **66**, 172

Propyne, 3-bromo-; (106-96-7), **66**, 86

1-PROPYNE, 3-CHLORO-, **65**, 47

2-Propyn-1-ol, **67**, 193

    silylation of, **66**, 3

2-Propyn-1-ol, 3-(trimethylsilyl)-; (5272-36-6), **66**, 7

Protodestannylation, **66**, 77, 78, 80

PULEGONE, (+): p-MENTH-4-(8)-EN-3-ONE, (R)-(+)-; CYCLOHEXANONE,
5-METHYL-2-(1-METHYLETHYLIDENE)-, (R)-; (89-82-7), **65**, 203, 215

4-Pyridinamine, N,N-dimethyl-, **65**, 12

Pyridine, **66**, 180, 182

Pyridine, 4-(dimethylamino)-, **65**, 12

Pyridine, hydrofluoride, **67**, 86

Pyridinium poly(hydrogen fluoride): Pyridine, hydrofluoride; (32001-55-1),
67, 86

2(1H)-Pyrimidone, tetrahydro-1,3-dimethyl-; (7226-23-5), **66**, 94

Pyrocatechol; (120-80-9), **66**, 184

Pyrophosphoric acid, disodium salt; (7758-16-9), **66**, 219

1 H-Pyrrole-2,5-dione, 1-phenyl-, **67**, 133

1-Pyrrolidinamine, N-(1-ethyl-2-methylpentylidene)-2-methoxymethyl)-,
[S-[R*,R*-(Z)]-, **65**, 183

1-Pyrrolidinamine, N-(1-ethylpropylidene)-2-(methoxymethyl)-, (S)-, **65**, 183

1-Pyrrolidinamine, 2-(methoxymethyl)-, (R)-(+)-; (72748-99-3), **65**, 173

1-Pyrrolidinamine, 2-(methoxymethyl)-, (S)-(-)-; (59983-30-0), **65**, 173, 183

Pyrrolidine, 2-(methoxymethyl)-, (S)-(+)-; (63126-47-6), **65**, 173

Pyrrolidine, 2-(methoxymethyl)-1-nitroso-, (S)-, **65**, 183

1-Pyrrolidinecarboxaldehyde, 2-(hydroxymethyl)-, (S)-(-)-; (55456-46-7),
65, 173

1-Pyrrolidinecarboxaldehyde, 2-(methoxymethyl)-, (S)-(-)-; (63126-45-4),
65, 173

2,5-PYRROLIDINEDIONE, 1-BROMO-, **65**, 243

2-Pyrrolidinemethanol, (S)-(+)-; (23356-96-9), **65**, 173

2-Pyrrolidinone, 1-vinyl-; (88-12-0), **66**, 149

Pyrrolo[3,4-c]pyrrole-1,3(2H,3aH)-dione, tetrahydro-2-phenyl-5-(phenylmethyl)-
cis-, **67**, 133

RADICAL CYCLIZATION, **66**, 75

Ramberg-Bäcklund reaction, **65**, 90

RAMP: (R)-1-Amino-2-methoxymethylpyrrolidine: 1-Pyrrolidinamine,
2-(methoxymethyl)-, (R)-; **65**, 173, 183

REDUCTION OF CARBOXYLIC ACIDS TO ALDEHYDES, **66**, 121

REDUCTIVE ANNULATION OF VINYL SULFONES, **67**, 163

Rhodium(1+), [[1,1'-binaphthalene]-2,2'-diylbis[diphenylphosphine]-P,P']-
[(1,2,5,6-η)-1,5-cyclooctadiene]-, stereoisomer, perchlorate, **67**, 33

Rhodium, di-μ-chlorobis[(1,2,5,6-η)-1,5-cyclooctadien]di-, **67**, 33

Rhodium on carbon, **67**, 187

RING EXPANSION, **65**, 17

    of cyclic ketoximes, **66**, 189

SALSOLIDINE, (-)-: ISOQUINOLINE, 1,2,3,4-TETRAHYDRO-6,7-DIMETHOXY-1-METHYL-, (S)- (493-48-1), **67**, 60

SAMP: (S)-1-Amino-2-methoxymethylpyrrolidine: 1-Pyrrolidinamine, 2-(methoxymethyl)-, (S)-; (59983-39-0), **65**, 173, 183

SELENOSULFONATION, **67**, 157

SILANE, 1,3-BUTADIYNE-1,4-DIYLBIS[TRIMETHYL-, **65**, 52

Silane, chloromethyldiphenyl-, **67**, 125

Silane, chloromethyltrimethyl-, **67**, 133

Silane, (3-chloropropyl)trimethyl-; (2344-83-4), **66**, 94

Silane, chlorotrimethyl-, **65**, 1, 6, 61; **66**, 6, 14, 16, 21, 44, 47

Silane, [1-cyclobutene-1,2-diylbis(oxy)]bis[trimethyl-, **65**, 17

Silane, (1-cyclobuten-1,2-ylenedioxy)bis[trimethyl-, **65**, 17

Silane, (1-cyclohexen-1-yloxy)trimethyl-, **67**, 141

Silane, [(1-ethoxycyclopropyl)oxy]trimethyl-; (27374-25-0), **66**, 51; **67**, 210

Silane, ethynyltrimethyl-, **65**, 52, 61

Silane, (3-iodopropyl)trimethyl-; (18135-48-3), **65**, 94

SILANE, (ISOPROPENYLOXY)TRIMETHYL-, **65, 1**

Silane, [(3-methoxy-1-methylene-2-propenyl)trimethyl-, **67**, 163

Silane, trichloro-, **67**, 20

SILANE, TRIMETHYL[(1-METHYLETHENYL)OXY]-, **65**, 1

Silane, trimethyl[[(4-methylphenyl)sulfonyl]ethynyl]-, **67**, 149

Silane, trimethyl (1-methyl-1,2-propadienyl)-; (74542-82-8), **66**, 7, 13

SILANE, TRIMETHYL(1-OXO-2-PROPENYL)-; (51023-60-0), **66**, 14, 21

Silane, trimethyl[(1-phenylethyl)oxy]-, **65**, 6, 12

Silane, trimethyl[(1-phenylvinyl)oxy]-, **65**, 6, 12

Silver nitrate, **66**, 111

 reaction with 1-trimethylsilyl-1-butene, **66**, 4

Silver(I) oxide; (20667-12-3), **66**, 111, 115

Silver perchlorate: Perchloric acid, silver(1+) salt, monohydrate;

 (14202-05-8), **67**, 33

Silver(I) trifluoroacetate; (2966-50-9), **66**, 115

 preparation, **66**, 111

Sodium; (7440-23-5), **66**, 76, 85, 96

Sodium amalgam, **66**, 96

Sodium bis(2-methoxyethoxy)aluminum hydride: Aluminate (1-), dihydrobis-

 (2-methoxyethanalato)-, sodium; (22722-98-1), **67**, 13

Sodium dicarbonyl(cyclopentadienyl)ferrate; (12152-20-4), **66**, 96, 107

Sodium ethoxide, **67**, 170

Sodium fluoride, aqueous, work-up for organoaluminum reaction, **66**, 188

Sodium hydride; (7646-69-7), **66**, 30, 32, 76, 79, 85, 109, 111, 114

Sodium iodide, **66**, 87

Sodium methoxide, **66**, 76

Sodium naphthalenide, **65**, 166

Sodium nitrite, **66**, 151

Sodium tungstate dihydrate: Tungstic acid, disodium salt, dihydrate;

 (10213-10-2), **65**, 166

Sonication, **67**, 133

 for reaction of 1,3-diaminopropane with alkali metals, **66**, 130

SPIRO[4.5]DECAN-1,4-DIONE; (39984-92-4), **65**, 17

Stannane, 1,2-ethenediylbis[dibutyl-, (E)-, **67**, 86

Stannane, tetrachloro-, **65**, 17

Stannane, tributyl-, **65**, 236

Stannane, tributylchloro-, **67**, 86

Stannane, tributylethynyl, **67**, 86

Stannane, vinylenebis[tributyl-, (E)-, **67**, 86

STETTER REACTION, **65**, 26

Succinic anhydride, **67**, 76

SUCCINIMIDE, N-BROMO-, **65**, 243

Sulfone, ethynyl p-tolyl, **67**, 149

2-Sulfonyloxaziridines, preparation, **66**, 207, 208

Sulfosalicylic acid spray for tlc plates, **66**, 216

Sulfur chloride, **65**, 159

Sulfur dichloride: Sulfur chloride; (10545-99-0), **66**, 159

Sulfur-diimide, dicarboxy-, dimethyl ester, **65**, 159

Sulfuryl chloride isocyanate, **65**, 135

Swern oxidation, **66**, 15, 18

Tartaric acid, dibenzoate, (-)- and (+)-, **67**, 20

2,3,4,6-Tetra-O-acetyl-α–D-glucopyranosyl bromide: Glucopyranosyl bromide tetraacetate, α–D-; α–D-glucopyranosyl bromide, 2,3,4,6-tetraacetate; (572-09-8), **65**, 236

Tetrabutylammonium fluoride; (429-41-4), **66**, 110, 111, 115

Tetrabutylammonium hydroxide; (2052-49-5), **66**, 214, 219

Tetrafluoroboric acid, **67**, 210

Tetrafluoroboric acid-diethyl ether complex; (67969-82-8), **66**, 97, 98, 100, 107

Tetrafluoroboric acid-dimethyl etherate: Borate(1-), tetrafluoro-, hydrogen,
compound with oxybis[methane] (1:1); (67969-83-9), **67**, 141

Tetraisopropyl titanate: Isopropyl alcohol, titanium (4+) salt;
2-Propanol, titanium (4+) salt; (546-68-9), **65**, 230; **67**, 180

Tetrakis(acetonitrile)palladium tetrafluoroborate; (21797-13-7), **66**, 52, 54, 59

Tetrakis(triphenylphosphine)palladium; (14221-01-3), **66**, 61, 62, 66, 68, 69, 74
67, 86, 98, 105, 114

4,5,4',5'-TETRAMETHOXY-1,1'-BIPHENYL-2,2'-DICARBOXALDEHYDE;
[1,1'-BIPHENYL]-2,2'-DICARBOXALDEHYDE, 4,4',5,5'-TETRAMETHOXY-;
(29237-14-7), **65**, 108

2,2,6,6-Tetramethylpiperidine: Piperidine, 2,2,6,6-tetramethyl-; (768-66-1),
67, 76

Tetramethyltin, **66**, 65

2,5,7,10-Tetraoxabicyclo[4.4.0]decane, cis-: p-Dioxino[2,3,-b]-p-dioxin,
hexahydro-; [1,4]-Dioxino[2,3-b]-1,4-dioxin, hexahydro;
(4362-05-4), **65**, 68

1,4,8,11-TETRATHIACYCLOTETRADECANE; (24194-61-4), **65**, 150

Tetravinyltin, **66**, 53, 55

Tetronic acids, **66**, 111, 112

Thiourea: Urea, thio-; (62-56-6), **65**, 150

Tin chloride, **65**, 17

Tin tetrachloride: Tin chloride; Stannane, tetrachloro-; (7646-78-8), **65**, 17

Titanium, triisopropoxymethyl-, **67**, 180

Tin, tetramethyl-, **66**, 55

Tin, tetravinyl-, **65**, 53, 55

Titanium chloride, **65**, 81, 6

Titanium tetrachloride: Titanium chloride; (7550-45-0), **65**, 81, 6; **66**, 8, 9

Toluene, *o*-iodo-; (615-37-2), **66**, 74

α–Toluenethiol, **65**, 215

o-Toluidine, 3-nitro-, **65**, 146

o-Tolyllithium, preparation, **66**, 69

p-Tolyl 2-(trimethylsilyl)ethynyl sulfone: Silane, trimethyl[[(4-methylphenyl)-
sulfonyl]ethynyl]-; (34452-56-7), **67**, 149

o-Tolylzinc chloride; (84109-17-1), **66**, 67, 74

Transesterification, **65**, 230

1,2,4-Triazine-3,5,6-tricarboxylic acid, triethyl ester; (74476-38-3), **66**, 149

1,2,4-Triazines, as Diels-Alder dienes, **66**, 147, 148

Tributylamine, **66**, 173, 176

Tributylethynylstannane: Stannane, tributylethynyl-; (994-89-8), **67**, 86

(Z)-3-Tributylstannylmethylene-4-isopropyl-1,1-cyclopentanedicarboxylic acid,
dimethyl ester, **66**, 77

Tributyltin chloride: Stannane, tributylchloro-; (1461-22-9), **67**, 86

Tributyltin hydride: Stannane, tributyl-; (688-73-3), **65**, 236; **66**, 77, 79, 86

2,3,6-Tricarboethoxypyridine, **66**, 145

Trichlorosilane: Silane, trichloro-; (10025-78-2), **67**, 20

Triethylamine, **66**, 15, 16, 174, 176, 185

Triethyl orthoacetate; (78-39-7), **66**, 22, 23, 28

Triethyl orthoformate: Orthoformic acid, triethyl ester; Ethane,
1,1',1"-[methylidynetris(oxy)]tris-; (122-51-0), **65**, 146

Triethyl phosphite: Phosphorous acid, triethyl ester; (122-52-1), **65**, 119, 108

Triethyl phosphonoacetate; (867-13-0), **66**, 220, 224

TRIETHYL 1,2,4-TRIAZINE-3,5,6-TRICARBOXYLATE; (74476-38-3),
66, 142, 144, 145, 149

Trifluoroacetic acid; (76-05-1), **66**, 111, 115, 136

Trifluoroacetic anhydride: Acetic acid, trifluoro-, anhydride; (407-25-0),
    65, 12; **66**, 153

[I,I-BIS(TRIFLUOROACETOXY)]IODOBENZENE, **66**, 132, 134, 136, 141

Trimethylamine, **66**, 29, 32

Trimethylamine, 1,1-dimethoxy-, **67**, 52

2,2,6-Trimethyl-1,3-dioxen-4-one; (5394-63-8), **66**, 195, 196, 202

3,4,5-Trimethyl-2,5-heptadien-4-ol: 1,5-Heptadien-4-ol, 3,4,5-trimethyl-;
    (64417-15-8), **65**, 42

Trimethyl ortho-4-bromobutanoate: Butane, 4-bromo-, 1,1,1-trimethoxy-;
    (55444-67-2), **67**, 193

Trimethyloxonium tetrafluoroborate: Oxonium, trimethyl-, tetrafluoroborate
    (1-); (420-37-1), **65**, 140

1-Trimethylsiloxycyclohexene: Silane, (1-cyclohexen-1-yloxy)trimethyl-;
    (6651-36-1), **67**, 141

1-Trimethylsiloxy-1-ethoxycyclopropane, **66**, 44

TRIMETHYLSILYLACETYLENE: SILANE, ETHYNYLTRIMETHYL-;
    (1066-54-2), **65**, 52, 61

3-Trimethylsilyl-2-propyn-1-ol, **66**, 1, 3

3-Trimethylsilyl-2-propyn-1-yl methanesulfonate, **66**, 2

Tripropylaluminum; (102-67-0), **66**, 186; 188, 193

Tris(tetrabutylammonium) hydrogen pyrophosphate trihydrate;
    (76947-02-9), **66**, 212, 219

Trithiane, sym-: S-Trithiane; 1,3,5-Trithiane; (291-21-4), **65**, 90

Tungstic acid, disodium salt, dihydrate, **65**, 166

ULLMANN REACTION, **65**, 108

2,5-Undecanedione; (7018-92-0), **65**, 26

2-Undecene, 2-methyl-, **67**, 125

Urea, thio-, **65**, 150

(S)-Valine, **66**, 153

Valinol: 1-Butanol, 2-amino-3-methyl-, (S)-; (2026-48-4), **67**, 52

Vilsmeier-Haack reagent, **66**, 121

Vibro-mixer, **65**, 52

Vinyl acetate: Acetic acid vinyl ester; Acetic acid ethenyl ester; (108-05-4), **65**, 135

VINYL CATION EQUIVALENT, **66**, 95

Vinyl cation synthons, **66**, 102

Vinyl radical cyclization, **66**, 81, 82

VINYLATION OF ENOLATES, **66**, 95, 104

Vinyllithium, **66**, 53, 55

N-Vinyl-2-pyrrolidone; (88-12-0), **66**, 145, 146, 149

Volatile reaction products, apparatus for collecting, **66**, 161, 162

Wittig rearrangement, **66**, 19

Zinc chloride; (7646-85-7), **66**, 43, 45, 51, 61, 62, 68, 69

Zinc, chloro(2-methylphenyl)-; (84109-17-1), **66**, 74

Zinc-copper couple: Copper, compound with zinc (1:1); (12019027-1), **67**, 98

ZINC HOMOENOLATE, **66**, 43

Zipper reaction, alkyne isomerization, **66**, 129